Issues and Challenges of Climate Change

Issues and Challenges of Climate Change

Editor: Andrew Hyman

www.callistoreference.com

Callisto Reference,
118-35 Queens Blvd., Suite 400,
Forest Hills, NY 11375, USA

Visit us on the World Wide Web at:
www.callistoreference.com

ISBN: 978-1-64116-735-2 (Hardback)

Cataloging-in-publication Data

Issues and challenges of climate change / edited by Andrew Hyman.
   p. cm.
Includes bibliographical references and index.
ISBN 978-1-64116-735-2
1. Climatic changes. 2. Climatology. 3. Climate change mitigation. 4. Food processing plants.
5. Food. I. Hyman, Andrew.
QC903 .C55 2023
363.738 74--dc23

TABLE OF CONTENTS

Preface

Climate change encompasses the global warming caused by humans such as increased greenhouse gas emissions, and the ensuing large-scale weather pattern adjustments. The effects of climate change are global in scope and unparalleled in scale. These effects range from changing weather patterns that jeopardize food production to rising sea levels that increase the risk of catastrophic flooding. There are various problems caused by climate change including species extinction, food insecurity, plastic waste crisis, air pollution, ocean acidification, and deforestation. These problems can be eliminated by investing in renewable substances, decreasing greenhouse gas emissions, reducing food waste, decreasing the usage of single use plastics, and putting more efforts in wildlife conservation. This book includes some of the vital pieces of work being conducted across the world, on various topics related to the issues and challenges of climate change. It will help the readers in keeping pace with the rapid changes in this area of study.

The information shared in this book is based on empirical researches made by veterans in this field of study. The elaborative information provided in this book will help the readers further their scope of knowledge leading to advancements in this field.

Finally, I would like to thank my fellow researchers who gave constructive feedback and my family members who supported me at every step of my research.

Editor

1

Water Resource Management Frameworks in Water-Related Adaptation to Climate Change

Godfrey Odongtoo, Denis Ssebuggwawo and Peter Okidi Lating

Contents

G. Odongtoo (✉)
Department of Computer Engineering, Busitema University, Tororo, Uganda

Department of Information Technology, Makerere University, Kampala, Uganda

D. Ssebuggwawo
Department of Computer Science, Kyambogo University, Kampala, Uganda
e-mail: dssebuggwawo@kyu.ac.ug

P. O. Lating
Department of Electrical and Computer Engineering, Makerere University, Kampala, Uganda
e-mail: plating@cedat.mak.ac.ug

Abstract

This chapter addresses the use of partial least squares–structural equation modeling (PLS-SEM) to determine the requirements for an effective development of water resource management frameworks. The authors developed a quantitative approach using Smart-PLS version 3 to reveal the views of different experts based on their experiences in water-related adaptation to climate change in the Lake Victoria Basin (LVB) in Uganda. A sample size of 152 was computed from a population size of 245 across the districts of Buikwe, Jinja, Mukono, Kampala, and Wakiso. The chapter aimed to determine the relationship among the availability of legal, regulatory, and administrative frameworks, public water investment, price and demand management, information requirements, coordination structures, and analytical frameworks and how they influence the development of water resource management frameworks. The findings revealed that the availability of legal, regulatory, and administrative frameworks, public water investment, price and demand management, information requirements, and coordination structures had significant and positive effects on the development of water resource management frameworks. Public water investment had the highest path coefficient ($\beta = 0.387$ and $p = 0.000$), thus indicating that it has the greatest influence on the development of water resource management frameworks. The R^2 value of the model was 0.714, which means that the five exogenous latent constructs collectively explained 71.4% of the variance in the development. The chapter suggests putting special emphasis on public water investment to achieve an effective development of water resource management frameworks. These findings can support the practitioners and decision makers engaged in water-related adaptation to climate change within the LVB and beyond.

Keywords

Climate change · Lake Victoria Basin · PLS-SEM · Water resource management frameworks

Introduction

Background

There is an increasing pressure on water resources in the Lake Victoria Basin (LVB) due to rapid population growth, increased urbanization and industrialization, uncontrolled environmental degradation, and pollution (Dauglas et al. 2014;

Bakibinga-ibembe et al. 2011). These pressures still remain as a big challenge to the sustainable management of water resources (Mongi et al. 2015). Moreover, the LVB remains to be an important water resource for five countries, namely, Uganda, Kenya, Tanzania, Rwanda, and Burundi. The LVB is the largest freshwater lake in Africa with a surface area of 68,800 km^2. It provides resources for fishing, agriculture, medicine, forestry, water transport, and other economic activities (Odongtoo et al. 2018). Its surrounding area is affected by increasing commercial activities and inadequate provision of sanitation services, among others (Okurut 2010; Wafula et al. 2014). These affect the landscape and water resources around the lake, making the water unsuitable for use.

According to Oyoo-Okoth et al. (2010), there is a combination of four analyzed heavy metals in the water samples of the LVB showing similar presence of heavy metal in the waters within the lake. This can be attributed to the fact that human activities, such as industrialization and agricultural practices, seriously contribute to the degradation and pollution of the environment, which adversely affect the water bodies, as noted by Devi et al. (2018). Increased human activities, such as poor land use, uncontrolled abstractions, and water pollution, greatly reduce the quantity and quality of the available water resources.

Globally, water resources are facing severe degradation due to pollution and inefficient water resource management strategies (Wang et al. 2015). It has been reported that water pollution causes approximately 14,000 deaths per day, mostly due to drinking water contamination caused by untreated sewage in developing countries (Devi et al. 2018). Rinawati et al. (2013) argued that these challenges create potential major threats to global biodiversity around water bodies, such as the LVB. The above authors further observed that this is true of the biodiversity around LVB as noted by Case (2006). The most recent statistics show that cities and small towns within the LVB are rapidly growing (Dauglas et al. 2014). The best examples of the growth pattern of towns and cities can be observed in Mwanza in Tanzania, Jinja and Port Bell in Uganda, and Kisumu in Kenya. Both population growth and urbanization have negative impacts on the quality and quantity of water resources. Some of these impacts include eutrophication and siltation, in addition to water pollution, as a result of increased runoff from bare lands as deforestation persists around the LVB (Ondieki 2015). Muhweezi (2014) also noted that biodiversity and ecosystem-specific goods and services around the LVB are likely to be adversely affected in the future by water pollution. It can be inferred that these challenges revolve around poor management of water resources. Failure to address such challenges will cause serious problem to the lives of many people who depend on the LVB.

Poor management of the scarce water resources can lead to climate change. The climate is continuously changing and is getting worse over time. Moreover, extreme weather conditions are occurring more frequently than before, thus negatively affecting the agricultural sector, consequently leading to food shortages. Farmers from different age groups acknowledge an increase in temperature, and they agree that the temperature is increasing with time (Mongi et al. 2015).

According to Leal et al. (2015), water is one of the essential resources for human being, and its availability has great impacts on the environmental, political, and economic situations. This means that water resources have to be managed well. Water resource management is the activity of planning, developing, distributing, and managing the optimum use of water resources (Okurut 2010). Ssozi et al. (2015) defined water resource management as the development of political, social, economic, and administrative systems to develop and manage water resources. According to the technical committee of the Global Water Partnership, Integrated Water Resources Management (IWRM) is a process that promotes the coordinated development and management of water, soil, and other related resources to maximize the resultant economic and social welfare in an equitable manner without compromising the sustainability of vital ecosystem (ITU-T 2010).

Analytical Framework

Water resource management has been recently developed due to the improvement in water technology, natural events, economic developments, and changing social needs. The current trend in water management changed from technical to a combination of technical and sociopolitical perspectives (Demetropoulou et al. 2010). This led to an increase in stakeholders and public participation. It is therefore necessary to establish a framework to guide practitioners and managers in water resource management. For managers and decision makers to conduct good analysis, it is necessary to use analytical frameworks to promote logical thinking in a systematic manner.

Adaptation and ICT Frameworks

According to Akoh et al. (2011), the frameworks of Information and Communication Technology (ICTs) in mitigating climate change and adaptation are still relatively new. Frameworks on adaptation to climate change can be considered in three ways: first, to emphasize the potential of ICTs to reduce vulnerability to climate change by building resilience; second, to focus on delivering different types of information needed to achieve effective climate change adaptation; and third, to emphasize a disaster risk management framework focusing on the reduction of community vulnerabilities and management and recovery from emergencies as they arise.

According to the 1992 Dublin Conference on water and the Rio de Janeiro Summit on sustainable development (White 2013), the four key principles adopted were as follows: first, freshwater is a finite and vulnerable resource essential to life, development of irrigation, industrial and transport sector, and the environment; second, water development and management should be based on a participatory approach, involving users, planners, and policy makers at all levels; three, women play a central part in the provision, management, and safeguarding of water; and fourth, water has an economic value in all its competing uses and should be recognized as an economic good.

Objective

The objective of this chapter was to apply a PLS-SEM to evaluate the requirements for the development of water resource management framework in water-related adaptation to climate change in Uganda's side of the LVB.

Methods

In evaluating the requirements for the development of water resource management framework in water-related adaptation to climate change, PLS-SEM was used. The proposed model was analyzed in two different stages: first, the model was composed of measurement models that define the relationship between latent indicators and their manifest variables, and second, a structural model showing the relationship between the manifest variables. A total of 24 factors which were obtained from the literature review were named as the observed variables and were divided into six groups: availability of legal, regulatory, and administrative frameworks, analytical framework, public water investment, information requirements, coordination structures, and pricing and demand management. These six groups were called exogenous latent constructs. Thus, an effective water resource management framework is influenced by the six major constructs.

Study Hypotheses

Based on the objectives of the research, the hypothesized model was developed. Quantitative data to test the hypothesis was collected from experts in water resource sectors and was subjected to Smart-PLS test. Some of the hypotheses passed the test and were therefore accepted, whereas the others that failed the test were rejected. Hypothesis 1 (H1): Availability of legal, regulatory, and administrative frameworks has a significant and positive effect on the development of water resource management frameworks. Hypothesis 2 (H2): Analytical framework has a significant and positive effect on the development of water resource management frameworks. Hypothesis 3 (H3): Public water investment has a significant and positive effect on the development of water resource management frameworks. Hypothesis 4 (H4): Information requirements has a significant and positive effect on the development of water resource management frameworks. Hypothesis 5 (H5): Coordination structures has a significant and positive effect on the development of water resource management frameworks. Hypothesis 6 (H6): Pricing and demand management has a significant and positive effect on the development of water resource management frameworks. This hypothesis design was adapted from Shahid et al. (2018).

Data Collection

The data collection procedures involved three important phases. In the first phase, preliminary variables were obtained to formulate hypotheses using the materials in the review of literature, such as books, journals, and conference materials. In phase two, a pilot study was conducted to ensure consistency and completeness to help modify the questionnaire. Lastly, the questionnaire survey was conducted to obtain the opinion of the respondents. This study procedure was adapted from Shahid et al. (2018). Data collection was performed in the Buikwe, Jinja, Mukono, Kampala, and Wakiso districts in Uganda. The above districts were chosen based on the fact that those areas heavily depend on the LVB and at the same time were immensely affected by the decline in water resources. The key stakeholders that were engaged in this chapter included employees from the LVB Commission, the LV Fishery Organization, District Environmental Officers, District Forestry Officers, NEMA, Ministry of Water and Environment, National and Regional Policymaking and Communication Organs, and key community leaders. To calculate the sample size, Slovin's formula was used, $n = \frac{N}{1 + Ne^2}$, where n denotes the sample size, N denotes the population size, and e denotes the error margin. The population size of 245 was used with a confidence interval of 95% and error margin of 5%, generating a sample size of 152.

The questionnaire was composed of two sections. Section one consisted of the respondents' personal information. Section two consisted of variables appropriately grouped into six categories based on the nature of the measurements/constructs: availability of legal, regulatory, and administrative frameworks (LRA), analytical frameworks (AF), public water investment (PW), information requirements (IR), coordination structures (CS), and pricing and demand management. The questionnaires were administered to different stakeholders with an experience of more than 5 years in the water resource sector, such as executives, managers, water engineers, and IT officers. During the 4 months of study, valuable opinions from experts were obtained and incorporated in the model.

Data Analysis

The hypothesized structural model was analyzed using Smart-PLS version 3. Smart-PLS has advantages over other regression-based methods: First, it is capable of evaluating several latent constructs with various manifest variables and is considered as the best technique for multivariate analysis (Shahid et al. 2018). Second, it is suitable for evaluating the constructs when the sample size is less than 200 in comparison with other SEM software, such as AMOS, which require a sample size of more than 200. In this specific chapter, the sample size was 152 and therefore justifies the use of Smart-PLS. Smart-PLS involves a two-step procedure, namely, evaluation of the outer measurement model and evaluation of the inner structural model (Henseler et al. 2009).

Results and Discussion

Respondents' Profile

Table 1 shows the demographic information of the respondents. The respondents were selected from a wide range of professionals engaged in the water resource sector. About 61.8% of the interviewed experts were male. Majority of the interviewed individuals were in the middle age group (30–39 years) (34.9%), followed by the young age group (20–29 years) (31.6%), (40–49) years were 23.7%. Those aged 50 years and above only accounted for 9.9% of the total population. This research is part of a bigger study conducted to develop a water resource management ICT model in the LVB.

Table 1 Demographic information

Response item	Count	Percent
Gender Distribution		
Female	58	38.2
Male	94	61.8
Total	152	100.0
Age Group		
20–29 years	48	31.6
30–39 years	53	34.9
40–49 years	36	23.7
50–59 years	14	9.2
Above 59 years	1	0.7
Total	152	100.0
Qualification		
PhD	4	2.6
Masters	68	44.7
Bachelors	67	44.1
Diploma	13	8.6
Total	152	100.0
Designation		
Manager/Administrator	33	21.7
Staff/Employee	100	65.8
Systems Administrator	8	5.3
Client/Customer	1	0.7
Total	152	100.0

Demographic Information of the Respondents

In Table 1, the highest percentage (44.7%) of the respondents have acquired master's degree, followed by bachelor's degree (44.1%), diploma (8.6%), and PhD (2.6%). The highest category of workers (65.8%) was staff/employee, followed by managers/administrators (21.7%), system administrators (5.3%), and ICT technicians (6.6%).

The level of education, the designation, the age, and the sex have significant effects on the perception of a person with regard to information. Older people tend to have more experiences, whereas highly educated ones display more in-depth knowledge on the subject matter of research. A low level of education has an impact on the use and adoption of technologies. The designation status is related more to the skills in the management of water resources. This promoted trust on the outcome of this study.

Evaluation of Outer Measurement Model

Outer Loading

In order to determine whether what was hypothesized was in line with the collected data, PLS-SEM was employed. PLS-SEM is very suitable for theory building and for examining the complex relationship of the models. The outer measurement model was used to calculate the reliability, internal consistency, and validity of the observed variables together with the unobserved variables (Gabriel et al. 2016). Consistency evaluations were based on a single observed and construct reliability tests, whereas convergent validity and discriminant validity were used for validity assessment. The observed variables with an outer loading of at least 0.7 are acceptable and should therefore be retained, whereas those with values less than 0.7 should be dropped (Shahid et al. 2018; Smith et al. 2014). From Table 2, the outer loadings of the retained variables ranged between 0.819 and 0.917 which are more than 0.7 and

Table 2 Construct reliability and validity

Main constructs	Item	Outer loading	T-statistics	CR	AVE
Availability of legal, regulatory, and administrative Frameworks	LRA2	0.799	12.766	0.812	0.683
	LR3	0.853	16.156		
Coordination structures	CS2	0.773	10.034	0.839	0.724
	CS4	0.922	40.008		
Information requirements	IR1	0.890	34.942	0.886	0.796
	IR2	43.681	43.681		
Public water investments	PWI1	13.500	13.5	0.742	0.591
	PWI3	9.463	9.463		
Price and demand managements	PDM2	14.359	14.359	0.810	0.681
	PDM3	19.859	19.859		

therefore valid. The variables, namely, collection and allocation, reuse of waste water, and water resource management, were dropped since their outer loading were below 0.7. Composite reliability (CR) was used for internal consistency evaluation in the construct reliability.

Average Variance (AVE)

To establish convergent validity on the construct, average variance extracted (AVE) is normally used. It is the grand mean value of the squared loadings of the indicators associated with the construct and is the sum of the squared loadings divided by the number of indicators. An AVE value of at least 0.50 shows that the construct explains more than half of the variance of its indicators (Hair et al. 2011). As shown in Table 2, the AVE values are greater than 0.5; therefore, both their convergent validity and internal validity are acceptable for this measurement model. In Table 3, the cross-loading of all observed variables are greater than the inter-correlations of other constructs in the model. The Fornell–Larcker criterion is a very good approach for assessing discriminant validity. It compares the square root of the AVE values with the latent variable correlations. The square root of each construct's AVE should be greater than its highest correlation with any other construct (Shahid et al. 2018). Therefore, these findings confirmed the cross-loading assessment standards which provided acceptable validation for the discriminant validity of the measurement model.

Using the t-test approach, some of the items that measure each construct were either retained or dropped, depending on whether the t-values are more or less than 2.76. The t-values must be greater than 2.76 for it to be retained as a measure for the variable (Hair et al. 2013; Shadfar and Malekmohammadi 2013). In Table 2, all the t-values of the retained items in measuring each construct are greater than 2.76.

Table 3 Discriminant validity for water resource management frameworks

	LRA	CS	IR	PDM	PDM
Availability of legal, regulatory, and administrative frameworks (LRA)	0.827				
Coordination structures (CS)	0.615	0.851			
Information requirements(IR)	0.092	0.135	0.892		
Price and demand management (PDM)	0.532	0.575	0.072	0.825	
Public water investment (PWI)	0.029	0.067	0.308	0.054	0.769

Table 4 Path coefficient

Hypothesized path	Standardized β	P-values
Availability of legal, regulatory, and administrative frameworks	0.244	0.000
Coordination structures (CS)	0.206	0.001
Information requirements(IR)	0.335	0.000
Price and demand management (PDM)	0.211	0.000
Public water investment (PWI)	0.387	0.000

Estimation of Path Coefficients (b) and P-values

The significance of the hypotheses was tested by calculating the *beta (β)* values. The
β values indicate the measure of the effect the variable has on the model. The higher
the values, the stronger the effect, and it is always <1. As shown in Table 4, both the
β and p values of every path in the hypothesized model were calculated.

In (H1), researchers hypothesized that availability of legal, regulatory, and administrative frameworks would significantly and positively influence the development of
water resource management frameworks. The findings in Table 4 and Fig. 1 confirm
that Availability of Legal, Regulatory, and Administrative Frameworks of water-related
factor significantly influenced the development of water resource management frameworks ($\beta = 0.244, p < 0.000$). Hence, H1 was excellently confirmed. Furthermore, the
findings from Table 4 and Fig. 1 hypothesized that the (H3) public water investment-
related factor positively influenced the development of water resource management
frameworks ($\beta = 0.387, p < 0.000$), showing that (H3) was effectively approved.
Information requirements (IR) significantly and positively influenced the development
of water resource management frameworks ($\beta = 0.335, p < 0.000$). Coordination
structure (CS)-related factor was positive and significant ($\beta = 0.206, p < 0.001$), thus
greatly supporting. The effect of price and demand management-related factor
($\beta = 0.211\ p < 0.000$) was also confirmed, thereby supporting (H6). The greater the
β coefficient, the stronger the effect of an exogenous latent construct on the endogenous
latent construct. Analytical frameworks were dropped from the list because they failed
both the t-test and the outer loading test.

Fig. 1 The graphical
representation of all path
coefficients of the water
resource management
frameworks

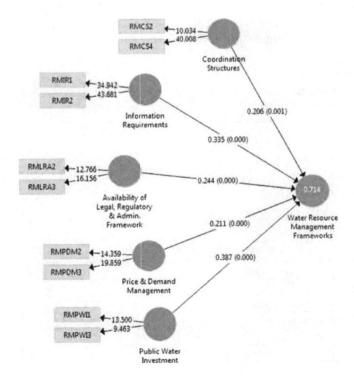

The R^2 value of the model was 0.714, which means that the five exogenous latent constructs collectively explained 71.4% of the variance in the development of water resource management frameworks. The study suggests putting special emphasis on the public water investment factor to achieve an effective development of water resource management frameworks. These findings can support the practitioners and decision makers engaged in water-related adaptation to climate change within the LVB and beyond.

Limitation of the Study

The limitation and constraint of the chapter came from the fact that it was not possible to visit all the countries of the East African Community due to logistical constraints. This will be part of the future works when the current logistical constraints are solved.

Business Benefit

In accordance with the complete analysis of the measurement models and structural model, some of the hypotheses were statistically significant and hence were accepted, whereas the others failed the analysis. The results of this chapter support a richer and accurate picture of the factors influencing the development of effective water resource management frameworks. It is therefore important to consider the availability of legal, regulatory, and administrative frameworks, coordination structures, information requirements, price and demand management, and public water investment factors in the development of effective water resource management frameworks. The information obtained from this chapter will also enrich the body of knowledge that will benefit other researchers in the field of water resource management.

Conclusion, Recommendation, and Future Research

Conclusion

The key contribution of this chapter was the empirical identification of the constructs that can influence the development of effective water resource management frameworks and also the investigation of the fundamental issues affecting the constructs observed by water experts in the LVB. The results of the chapter revealed that availability of legal, regulatory, and administrative frameworks, coordination structures, information requirements, price and demand management, and public water investment had significant positive effects on the development of water resource management frameworks. This therefore suggests that emphasis should be put on the above variables. The final results of the partial least squares–structural equation modeling revealed that public water investment had the highest path coefficient ($\beta = 0.387$), thus indicating a major influence on the water resource management frameworks. Therefore, water resource managers should pay more attention to public water investment factors during the development of water resource management frameworks.

Recommendation

This chapter recommends that for a successful development of water resource management frameworks, emphasis should be put on availability of legal, regulatory, and administrative frameworks, coordination structures, information requirements, price and demand management, and public water investment. Since public water investment had the highest path coefficient ($\beta = 0.387$) with a major overall influence on the development of water resource management frameworks, more emphasis should be put on it.

Lesson Learned and Future Studies

This chapter revealed that majority of the people engaged in different activities around the LVB are unaware of the dangers posed by their economic activities. Moreover, the local leaders lack effective and efficient means of disseminating and sharing information on the LVB and the impact of human activities on ecosystems, biodiversity, and water resources.

Future Studies

The areas for further research may focus on the study of how effective water resource management frameworks influence the design of an effective water resource management ICT model for an integrated water resource management of the LVB .

References

Akoh B, Livia B, Perry J-E, Creech H, Karami J, Echeverria D, ... Gass P (2011) Africa transformation-ready: the strategic application of information and communication technologies to climate change adaptation in Africa. Prepared for the African Development Bank, the World Bank and the African Union, Bristol

Bakibinga-ibembe DJ, Said AV, Mungai WN (2011) Environmental laws and policies related to periodic flooding and sedimentation in the Lake Victoria Basin of East Africa. Afr J Environ Sci Technol 5(5):367–380

Case (2006) Climate change impacts on East Africa. A review of the scientific literature. WWF-World Wide Fund for Nature (formerly World Wildlife Fund), Gland

Dauglas WJ, Hongtao W, Fengting L (2014) Impacts of population growth and economic development on water quality of a lake: case study of Lake Victoria Kenya water. Environ Sci Pollut Res 21:5737–5746. https://doi.org/10.1007/s11356-014-2524-5

Demetropoulou L, Nikolaos N, Papadoulakis V, Tsakiris K, Koussouris T, Kalogerakis N, Koukaras K (2010) Water framework directive implementation in Greece: introducing participation in water governance – the case of the Evrotas River Basin Management Plan. Environ Policy Gov 20:336–349

Devi SP, Jothi S, Devi A (2018) Data mining case study for water quality prediction using R tool. Int J Sci Res Comput Sci Eng Inform Technol 3(1):262–269

Gabriel C, Christian MR, Jörg H, José L (2016) Prediction-oriented modeling in business research by means of PLS path-modeling. J Bus Res 69(10):4545–4551

Hair JF, Ringle C, Sarstedt M (2011) PLS-SEM: indeed a silver bullet. Market Theory Pract 19:139–151

Hair FJ, Ringle MC, Sarstedt M (2013) Partial least square structural equation modelling: rigorous application, better results and higher acceptance. Long Range Plan 46:1–12

Henseler J, Ringle C, Sinkovics R (2009) The use of partial least squares path modeling in international marketing. Adv Int Market 20:277–319

Leal FW, Trincheria DJ, Vogt J (2015) Towards sustainable water use: experiences from the projects AFRHINET and Baltic flows. In: Leal Filho W, Sümer V (eds) Sustainable water use and management, green energy and technology. Springer, Cham

Mauree V (2010) ICT as an enabler for smart water management. ITU-T Technology Watch Report, ITU Telecommunication Standardization Bereau, p 23

Mongi HJ, Mvuma AN, Kucel S, Tenge AJ, Gabriel M (2015) Accessibility and utilization of mobile phones for governance of water resources in the Lake Victoria Basin: constraints and opportunities in Tanzania. Afr J Environ Sci Technol 9(5):438–450

Muhweezi AB (2014) An overview of climate change and biodiversity in Uganda. African and Latin American Resilience to Climate Change Project. https://doi.org/AID-OAA-TO-11-00064

Odongtoo G, Ssebugwawo D, Okidi LP (2018) Factors affecting communication and information sharing for water resource management in Lake Victoria. In: Handbook of climate change and biodiversity. Springer, Cham, pp 211–222

Okurut T (2010) Integrated environmental protection approaches: strengthening the role of water supply operators in resource conservation. In: 15th international African water congress and exhibition commonwealth resort, Munyonyo, Kampala, Uganda

Ondieki MC (2015) Aspects of water quality and water pollution control in vulnerable environments. J Geogr Earth Sci 3(2):83–93

Oyoo-Okoth E, Wim A, Osano O, Kraak MHS, Ngure V, Makwali J, Orina PS (2010) Use of the fish endoparasite ligula intestinalis (L., 1758) in an intermediate cyprinid host (Rastreneobola argentea) for biomonitoring heavy metal contamination in Lake Victoria, Kenya. Lakes Reserv Res Manag 15:63–73

Rinawati F, Stein K, Lindner A (2013) Climate change impacts on biodiversity – the setting of a lingering global crisis. Biodiversity 5:114–112

Shadfar S, Malekmohammadi I (2013) Application of structural equation modeling (SEM) in restructuring state intervention strategies toward paddy production development. Int J Acad Res Bus Soc Sci 3(12):576

Shahid H, Fangwei Z, Ahmed FS, Ali Z, Muhammad SS (2018) Structural equation model for evaluating factors affecting quality of social infrastructure projects. Open Access J Sustain 10:1415

Smith RG, Davis AS, Jordan NR, Atwood LW, Daly AB, Grandy AS, ... Yannarell AC (2014) Structural equation modeling facilitates transdisciplinary research on agriculture and climate change. Crop Sci 54:475–483

Ssozi F, Blake E, Rivett U (2015) Designing for sustainability: involving communities in developing ICT interventions to support water resource management. https://doi.org/978-1-905824-51-9

Wafula JD, Wang H, Li F (2014) Impacts of population growth and economic development on water quality of a lake: case study of Lake Victoria Kenya water (Garrigues P, ed). Springer, Berlin/Heidelberg

Wang Z, Song H, Watkins DW, Ong KG, Xue P, Yang Q, Shi X (2015) Cyber-physical systems for water sustainability: challenges and opportunities. IEEE Commun Mag 53(5):216–222

White C (2013) Integrated water resources management: what is it and why is it used? Retrieved from https://globalwaterforum.org/2013/06/10/integrated-water-resources-management-what-is-it-and-why-is-it

2

Sorghum Farmers' Climate Change Adaptation Strategies in the Semiarid Region of Cameroon

Salé Abou, Madi Ali, Anselme Wakponou and Armel Sambo

Contents

Abstract

This chapter deals with the problem of sorghum farmers' adaptation to climate change in the semiarid region of Cameroon. Its general objective is to compare the various adaptation strategies' typologies and to characterize the sorghum farmers' adaptation strategies on the basis of the suitable one. The stratified

S. Abou (✉) · M. Ali
National Advanced School of Engineering of Maroua (ENSPM), The University of Maroua, Maroua, Cameroon

A. Wakponou
Faculty of Arts, Letters and Human Sciences (FALSH), The University of Ngaoundéré, Ngaoundéré, Cameroon

A. Sambo
Faculty of Arts, Letters and Human Sciences (FALSH), The University of Maroua, Maroua, Cameroon

random sampling method was used to select the sites, which consist of twenty (20) villages, and the sample, which consists of six hundred (600) farm household heads. After conducting focus-groups in ten villages and interviews with resource persons, the primary data were collected using a semi-open survey questionnaire. It appears that the poor spatiotemporal distribution of rains and the drought constitute, respectively, the main climate hazard and the main water risk that farmers are dealing with; the farmers are vulnerable to climate change because the adaptation strategies used are mostly traditional, their adoption rates are very low, and the use of efficient adaptation strategies (irrigation, improved crop varieties) is almost unknown. The characterization of the adaptation strategies used shows that they are more complex than most authors who have established the typologies thought. It comes out that improving the resilience of these sorghum farmers absolutely requires the improvement of their basic socioeconomic conditions.

Keywords

Semiarid region · Sorghum farmers · Climate change · Climate hazard · Water risk · Adaptation strategies

Introduction

Farmers in the semiarid regions of Africa, to which belongs the Diamaré division in the Far North Cameroon, are among the most vulnerable to water constraints caused by climate variability during the 1960s and 1970s. This vulnerability has its origin in their essentially rain-fed agriculture, their unfavorable socioeconomic characteristics, and their very fragile ecosystem.

According to Borton and Nicholds (1994), of all-natural hazards, droughts are the ones with the greatest economic impact, and affecting the greatest number of people. In the Diamaré division, as in the whole semiarid zone of Cameroon, water constraints, particularly droughts and floods, have had a negative impact on cereals' production, especially sorghums, which constitute the basic food of the population. According to L'hôte (2000), the period called "Drought in the Sahel" was an agronomic disaster for the entire region. Similarly, the results of the simulations carried out by Blanc (2012), compared to a reference without climate change, indicate that sorghum yields could decrease by around −47% to −7% by 2,100 in this region. Faced with this situation, a wide variety of adaptation strategies emanating from both farmers and agricultural research has been made available to sorghum farmers, but adoption rate remains low as everywhere else in the African semiarid zones (Yesuf et al. 2008; Leary Kulkarmi and Seipt 2007).

In order to better understand the main orientations of these various adaptation strategies, a variety of typologies has been previously established by some authors (Dingkuhn 2009; Nhemachena and Hassan 2007; Jouve 2010; then Fabre 2010); but a comparison between adaptation strategies on the basis of these typologies remains difficult because of the diversity of analysis' angles used by the authors. For this reason, it seems better to identify the main similarities and differences between the various typologies, and then to characterize the sorghum farmers' adaptation strategies on the basis of the most suitable typology. The sorghum farmers' adaptation strategies' characterization could allow researchers as well as policy-makers to better reorient research priorities and policies in order to improve farmers' resilience.

The Diamaré division located in the Far-North region of Cameroon (Fig. 1), between 10° and 11° North latitude (10°30'00'') and 14° and 15° East longitude (14° 30'00''), constituted the focus area. The climate is Sudano-Sahelian in its southern part and Sahelo-Sudanian in its northern part, all characterized by a long dry season (7–9 months), and a short rainy season. Agriculture (rainy season, dry season), animal husbandry, fishing, trade, and crafts are the main activities of these populations.

The information has been collected through directed interviews with some resource people (researchers, patriarchs, heads of technical services), and then focus groups and a survey questionnaire submitted to six hundred (600) household heads. The descriptive and inferential statistics (frequencies, percentages, Principal Component Analysis, Kaiser-Meyer-Olkin/KMO test) from the SPSS statistical software have been used to analyze the information gathered.

Socioeconomic Characteristics of the Sorghum Farmers

In general, agriculture and livestock are the main activities of the farmers, and the Diamaré division is one of the three divisions most exposed to food insecurity in the region. Priority is given to cereals in terms of land mobilization and work (CEDC 2010), and sorghum (rainy and dry seasons) constitute the staple food of the populations. This agriculture is essentially characterized by the practice of polyculture (93.80%) and self-consumption agriculture (79%), the small size of the sown areas (100%<10 ha, with areas varying between 0,5 ha and 1 ha for sorghum), the low quantity of agricultural inputs used (FAO 2009), the poor access to agricultural supervision (51.50%), and to credits (43.50%). Cotton and onion are the only export crops. In order to ensure their daily survival, these farmers multiply income-generating activities (65.50%), which reflects the inability of agricultural activities to meet their food needs, and therefore their high vulnerability. The household heads are mostly men, most of them aged between 35 and 54 years (FAO 2009), with an average age of 48 years in Diamaré. The average household size is around 9 people, and seems to be high compared to a regional average of 7 and a national average of 5.7. The school enrollment rate, which is 57%, is the lowest in the country, with 39.30% having reached primary, 17% secondary, and less than 1% (0.4%) higher education. Health and school infrastructures are among the least fortunate in the country.

Climate Hazards and Sorghum Farmers' Adaptation Strategies

Table 1 summarizes the climate hazards imposed by the climate change to the rainy and dry seasons' sorghum production.

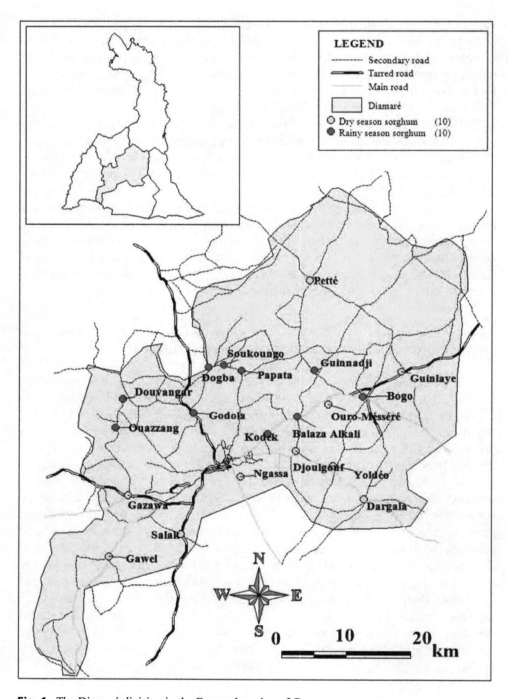

Fig. 1 The Diamaré division in the Far-north region of Cameroon

Table 1 Climate hazards perceived by rainy and dry seasons sorghums' farmers

Climate hazards	Rainy season sorghum		Dry season sorghum	
	Total	%	Total	%
Late or early arrival of rains	300	100	297	99
Early cessation of rains	300	100	293	97,67
Poor spatial rainfall distribution	299	99,67	299	99,67
More frequent and long dry spells	300	100	300	100
Flooding of crops	300	100	294	98
General decline in the total amount of rainfall	299	99,67	296	98,67
Rapid drying of water sources (ponds, rivers, lakes, wells)	–	–	300	100
Rapid drying and hardening of soils	–	–	292	97,33
Light rains at the beginning of the rainy season	–	–	297	99
Absence of heavy rains at the end of the rainy season	–	–	297	99
No haze during cold season	–	–	299	99,67

Most of the climate hazards listed relate to the poor spatiotemporal distribution of rains, while most of them are linked to drought, and not to excess water (floods); that means the poor spatiotemporal distribution of rains is the main climate hazard while drought is the main water risk limiting agricultural production both during the rainy and dry seasons in the area. This observation confirms the results obtained by Chédé (2012), Gnanglé et al. (2012), then Agossou et al. (2012).

Likewise, it emerges that all the three known forms of drought are represented here:

- Meteorological drought (late arrival of the rains, early start of the rains, dry spells, decrease in the amount of rains, light rains at the start of the rainy season, light rains at the end of the rainy season)
- Agricultural drought (rapid drying and hardening of soils)
- Hydrological drought (rapid drying up of ponds and rivers)

According to FAO and National Drought Mitigation Center (2008), when all of these three forms of drought rage somewhere, automatically, the socioeconomic drought which is their logical consequence, will also rage there; and that is noticeable through the socioeconomic characteristics of the sorghum farmers described in the previous paragraph.

One can also remark the high rate of climate hazards' perception by the farmers, which could reflect both the extent and the severity of these constraints, but also a good perception of the phenomena by farmers who control their physical environment. Indeed, Arodokoun (2011) considers that, in general, peasant communities which maintain close links with their environment have a perfect knowledge of the climate, its manifestations and the changes that have occurred. This situation could also be justified by the fact that according to Nhemachena and Hassan (2007), climate change has very negative effects on the poorest households which have the least capacity to adapt to changing climatic conditions.

It appears also that all the climate hazards affecting rain-fed sorghums also affect dry season sorghums; which reflects the strong dependence of dry season sorghum on meteoric water. On the other hand, the multiplicity of climate hazards affecting dry season sorghum can be explained by the diversity of the water resources (meteoric, surface, underground) essential to its production.

The adaptation strategies in use by sorghum farmers in the face of these climate hazards and water risks are summarized in Table 2.

The analysis of the nature of the adaptation strategies used by sorghum farmers in the face of climate hazards and water risks could lead to the following remarks:

- All the adaptation strategies used by sorghum farmers aim to compensate for either the poor distribution of rains, the droughts (meteorological, edaphic, hydrological, socioeconomic), or to both of the two types of constraints.
- An overwhelming majority of these adaptation strategies have been adopted to cope with meteorological drought, which is the main form of drought faced by sorghum farmers; it is followed by edaphic drought, then hydrological drought.

Table 2 Nature and frequency of adoption of the adaptation strategies used by sorghum farmers

Adaptation strategies	Rainy season sorghum		Dry season sorghum	
	Total	(%)	Total	(%)
Sowing early matured varieties	131	43,67	175	**58,33**
Sowing or transplanting early	178	**59,33**	139	**46,33**
Sowing of drought resistant crops varieties	178	**59,33**	194	**64,67**
Diversification of crops varieties	94	31,33	182	**60,67**
Changing of crops or crops' varieties	105	35	25	08,33
Labor of plots and mounding of plants	234	**78**	96	32
Temporary or permanent transfer of crops	170	**56,67**	30	10
Making of racks or bunds	103	34,33	203	**67,67**
Nursery organic or inorganic fertilizer input	271	**90,33**	82	27,33
Diversification of income-generating activities	195	65	141	47
Crops diversification	268	**89,33**	272	**90,67**
Multiplication of weeding	123	41	20	06,67
Sowing of molten seed holes or dried plants	166	**55,33**	05	01,67
Rocky bunds	05	01,67	–	–
Late transplanting	–	–	125	**41,67**
Deepening piles	–	–	129	43
Purchase or request of nurseries	–	–	104	34,67
Scaling of nurseries over the time	–	–	203	**67,67**
Organic or inorganic fertilization of nurseries	–	–	107	35,67
Cleaning and deepening of ponds	–	–	131	**60,33**
Water research over long distances	–	–	95	31,67
Fertilization of transplanting water	–	–	06	02

- Despite the identification of floods as another main water risk by sorghum farmers, no adaptation strategy has apparently been adopted by them to deal with it.
- The adaptation strategies used are traditional in their majority, and those recognized as efficient such as the irrigation or the use of improved crops' varieties are almost unused.
- Despite the traditional nature of the adaptation strategies used, their adoption rates by farmers are low in the majority of cases.

The synthesis of these preceding remarks conducts to the following conclusions:

- The poor spatiotemporal distribution of rains and the drought (especially meteorological), constitute, respectively, the main climate hazard and the main water risk faced by farmers in the area; this has been previously confirmed by several researchers including Batterbury and Mortimore (2013), then Mortimore and Adams (2000), who consider that all the Sahelian farmers' problems correspond to a group of "five (5) crises of the Sahelian orthodoxy," to which they are trying to provide solutions, the main one of which is represented by drought.
- Sorghum farmers in particular and farmers in general from the region are very vulnerable because they do not really adapt to but they simply cope with the climate change; this confirms the observation made by the IPCC (2014), Sissoko et al. (2010), OECD (2010), then Leary et al. (2007), according to whom Sahelian farmers do not adapt to but simply cope with climate change; these sorghum farmers' maladaptation (lack of adaptation) can be perceived or explained from the previous paragraph 1 through the practice of self-consumption agriculture, the small size of the sown areas, the low quantity of agricultural inputs used, the poor access to agricultural supervision and to credits, the multiplication of income-generating activities, and the weak school enrollment rate. Contrary to this result, authors like Jouve (2010) and Batterbury and Forsyth (1999) find that in fact Sahelian farmers adapt to climate change. This duality of contradictory observations could be justified either by the different use of the concept of adaptation, or by the difference in the results obtained in different contexts.

Sahelian Farmers Adaptation Strategies' Typologies

On the whole, the farmers' adaptation strategies in the dry regions (semiarid, arid) can be classified both on the basis of farmers' practices, which takes into account all the actions undertaken by them to ensure their survival, and on the basis of the agricultural research, which is only interested in crop management. According to these two angles of analysis, the typologies could be grouped into three main categories:

- The first category of typologies, proposed by Nhemachena and Hassan (2007), Jouve (2010), Ngigi (2009), Diarra (2008), Fabre (2010), then Batterbury and Mortimore (2013).
- The second category of typologies brings together those proposed by Batterbury and Forsyth (1999), then Ngigi (2009).
- The third category of typologies, which corresponds to that proposed by Dingkuhn (2009).

First Category of Typologies

The first category of typologies is that proposed by Nhemachena and Hassan (2007), Jouve (2010), Ngigi (2009), Diarra (2008), Fabre (2010), then Batterbury and Mortimore (2013). These typologies distinguish adaptation strategies which consist of facing risks from those focusing on avoiding risks.

In general, Nhemachena and Hassan (2007) estimate that there are roughly two types of adaptation strategies in agricultural production systems:

1. The strategies which are based on "confronting water risks," essentially based on increasing diversification which corresponds to the adoption of production activities tolerant to drought and/or resistant to thermal stress, as well as activities that relate to efficient management that value the quantity of water available and ambient temperatures among other factors.
2. The strategies which are based on "eviction of water risks," essentially based on crop management practices, and which consist in avoiding that the critical stages of plant growth do not coincide with bad climatic conditions such as inter-season droughts (modification of the crop cycle and modification of sowing and harvest dates).

Jouve (2010), Ngigi (2009), and Diarra (2008), based on the subjective assessment of risks and vulnerability, have grouped the farmers' adaptation strategies into three (3) categories:

1. The "pre-risk" or preventive management options (prevention strategies) or before risks, such as the choice between risk-tolerant varieties, investment in water management, and diversification of survival and agriculture, well before the arrival of the growing season.
2. The "intra-season" adjustment of crops and resources management options in response to constantly changing specific climatic shocks, also called "adaptive methods" or "mitigation strategies"; these are peasant innovations put in place to adapt to climate change and to make the best use of rainwater resources.
3. The "post-risk" management options or "palliative methods" or "adjustment strategies," which minimize the impacts of adverse climate shocks; they seek to mitigate the effects of climatic risks which particularly affect poor rural

populations; these methods, based on the establishment of insurance systems, aim to stabilize farmers' incomes and avoid their indebtedness and decapitalization during bad years; they are intended to alleviate the impact of the event when it happened.

The comparison between these two typologies indicates that they are identical: the strategies based on "confronting water risks" proposed by Nhemachena and Hassan (2007) correspond to the intra-season adjustment and post-risk" management options in the typology proposed by Jouve (2010), Ngigi (2009), and Diarra (2008), while "crowding out water risks" corresponds to "pre-risk" management options.

Batterbury and Mortimore (2013) estimate that the Sahelian farmers' adaptation strategies correspond to the five crises of Sahelian orthodoxy:

- The management of rainfall by farmers each year
- The integration of agriculture and animal husbandry
- The conservation management of biodiversity
- The intensive and sustainable soil management
- The diversification of livelihoods

Jouve (2010) and Fabre (2010), for their part, believe that sahelian farmers' adaptation strategies to climate variability can be split into three (3) groups:

- The choice of crops (species, varieties)
- The modification of practices (irrigation and drainage, polyculture, modification of the cropping calendar)
- The modification of sources of income (crafts, livestock, trade, etc.)

Reconciling the last two typologies also indicates that they are similar insofar as the management of rainfall, the integration of agriculture and animal husbandry, and the intensive and sustainable management of soil suggested by Batterbury and Mortimore (2013) corresponds to the modification of practices suggested by Jouve (2010) and Fabre (2010). Similarly, the conservation management of biodiversity corresponds to the choice of crops, while the diversification of livelihoods corresponds to the modification of sources of income.

Finally, the comparison between the two previous typologies and the two last ones indicates that they are in fact similar. The "increase in diversification" proposed by the first two typologies corresponds to the "modification of the sources of income and the practices" proposed by the two last ones, while the crop "management practices" correspond to the "choice of crops"; this amounts to saying that in terms of farmers' practices, the four typologies are identical.

The particularity of this first category of typologies is that they are based principally on the natural agro-pastoral resources (water, soils, crops, animals) spatiotemporal management.

Second Category of Typologies

The second category of typologies includes those proposed by Batterbury and Forsyth (1999), then Ngigi (2009). These typologies distinguish "adaptive processes" which are long and medium terms adaptation strategies, from "adaptive strategies" which are short term actions intended to ensure the survival of farmers.

Batterbury and Forsyth (1999) find that farmers' adaptation strategies can be divided into two categories:

- The "adaptive processes," which generally call for a spatial extension of activities outside the locality, in order to reduce the pressure on local resources; it is an appropriate strategy for communities in dry regions where diversification is the main response to drought or crop failure; adaptation processes are long-term transitions that change the configuration of relationships between a community and its resources; each transition has several components, and adoption and the form that transition takes depends on several factors of change, such as farmers' knowledge, the biophysical environment (especially precipitation and soil), and availability of the work force; for this reason, each transition will be relatively unique, thus reflecting the interactions between farmers, their institutions, their economic policy, and their environment.
- The "adaptive strategies" are short-term practices, adopted in response to sudden shocks or difficulties in accessing resources.

Ngigi (2009) differentiate adaptation strategies to climate change between:

- "Adaptation itself" or "adaptation strategies," which constitutes a change in response to changing climatic parameters and
- "Coping mechanisms" or "coping strategies," generally in the short term.

The comparison between these two typologies shows that they are also similar. "Adaptation itself" corresponds to "adaptive processes," while "coping strategies" correspond to "adaptive strategies."

This category of typologies is characterized by the fact that tries to differentiate the farmers' adaptation strategies, either on the long- and medium-terms use of agricultural or nonagricultural strategies to truly adapt climate change, or on the short-term use of agricultural or nonagricultural strategies to cope with climate change (without really adapting to it); in other words, it differentiates farmers' adaptation strategies based on whether they are genuinely adapting in the long- to medium-terms, or whether they are simply coping in a short-term with climate change to just ensure their survival.

Third Category of Typologies

The typology proposed by Dingkuhn (2009), corresponds to the third category, and groups adaptation strategies according to the agricultural research fields, into four very distinct types:

- Genetic adaptation (drought resistant varieties, early varieties)
- Agronomic adaptation (all the strategies linked to the management of crops, which are the most numerous)
- Geographic adaptation (temporary or permanent change in cultures)
- Temporal adaptation (early sowing, late sowing, staggering of nurseries)

The particularity of this category of typologies is that it is not interested in the climate hazards or risks that farmers are facing (first category of typologies), nor in the strength or weaknesses of the adaptation strategies in use (second category of typologies), but simply to the scientific field to which each of these adaptation strategies relates.

The comparison between all the previous typologies and the current one proposed by Dingkuhn (2009) indicates that one could very well transpose the genetic, agronomic, geographic, and temporal adaptations suggested by this typology in these ones; however, the socioeconomic adaptation taken into account by these typologies does not exist in the one proposed by Dingkuhn (2009), and this could be explained by the fact that this typology is essentially concerned with crop management and not with farmers practices in their whole.

After a careful analysis of the typologies' categories and the sorghum farmers' adaptation strategies, it appears that it is appropriate to characterize them on the basis of the first category of typologies because both (category of typologies, adaptation strategies) are oriented towards the agro-pastoral natural resources' management; on the other hand, while the second category of typologies requires the strategies to be monitored over time in order to assess their effectiveness, the third category is more descriptive and does not allow to grasp easily the real objectives targeted by sorghum farmers.

Characterization of Sorghum Farmers' Adaptation Strategies

The adaptation strategies were grouped according to crops, so first, the adequacy of the rainy season sorghum farmers' adaptation strategies has been tested using the KMO test. The results are listed in Table 3.

All the KMO values taken by the different adaptation strategies being greater than 0.49, all these strategies have been used in the test.

Table 3 Results of the KMO sample adequacy test applied to the rainy season sorghum farmers' adaptation strategies

Adaptation strategies	Codes	KMO values
Sowing early matured varieties	SEVARPRE	0.834
Sowing or transplanting early	SEMISSEC	0.743
Sowing of drought resistant crops varieties	SEMVARES	0.817
Diversification of crops varieties	DIVARCUL	0.896
Change of crops or crops varieties	CHASPVA	0.671
Labor of plots and mounding of plants	LABBUTPL	0.815
Temporary or permanent transfer of crops	MUTEDECU	0.858
Making of racks	CONFCAS	0.925
Organic fertilizer input	FUMORGA	0.849
Diversification of income-generating activities	DIVACGER	0.854
Crops' diversification	DIVERCUL	0.850
Multiplication of weeding	MULTSARC	0.867
Sowing of molten seed holes or dried plants	RESREPIQ	0.845

The eigenvalues of the various factors from the PCA results reveal the existence of two (2) main factors, which explain 53.64% (>49%) of the total variation of the adaptation strategies, in accordance with the KMO rule (Fig. 2 and Table 4).

Loading these adaptation strategies according to the two main factors gave the results mentioned in Table 5.

Factor 1 brings together the adaptation strategies "sowing of early matured varieties," "diversification of crop varieties," "change of crops or crop varieties," "labor of plots and mounding of plants," "temporary or permanent transfer of crops," "organic or mineral fertilizer input," and "sowing of melted seed holes or dried plants." This factor can be called "Adaptation to climate hazards through efficient management of natural resources (soil, water, crops)." This factor can be interpreted as the decision-making by farmers to continue to carry out agricultural activities despite the risks, and corresponds in fact to the adaptation strategy by "**confronting climate hazards and water risks**" suggested by the first category of typologies.

Factor 2 groups together the adaptation strategies "sowing or transplanting early," "sowing of drought resistant crops varieties," "making of racks," "diversification of income-generating activities," "crop diversification," and "multiplication of weeding." This factor brings together adaptation strategies whose main objective is to avoid water risks.

It therefore corresponds to all the activities carried out by farmers with the aim of minimizing climate hazards and water risks and their impacts; and for that, it corresponds well to the adaptation strategy by "**eviction or minimization a priori of climate hazards and water risks**" suggested by the first category of typologies.

It finally emerges from this analysis that all the rainy season sorghum farmers' adaptation strategies correspond very well to the typology proposed by the first category of typologies' authors, namely adaptation by "**confronting water risks**" and adaptation by "**a priori eviction or minimization of water risks**".

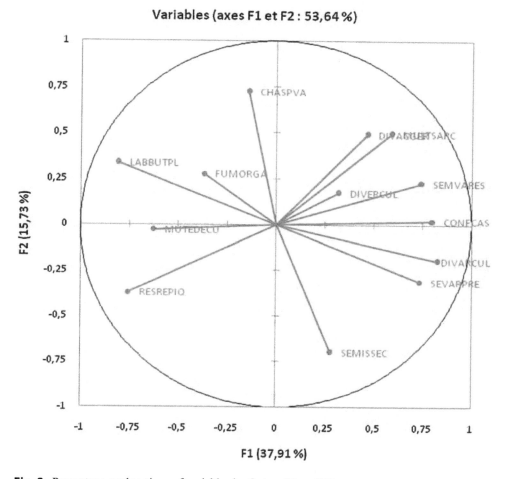

Fig. 2 Percentage explanations of variables by factors F1 and F2

Table 4 Variability explained by factor

Factors	Variation	%	Cumulative %
F1	4.928	37.908	37.908
F2	2.045	15.731	**53.64**

The test of the adequacy of the dry season sorghum farmers' adaptation strategies using the KMO test gave the results mentioned in Table 6.

All the KMO values taken by the adaptation strategies are greater than 0.49, except that of the "late transplanting" strategy, which will not be used in the KMO test.

The eigenvalues of the different factors from the PCA results reveal the existence of five (5) main factors, which explain 53.612% (>49%) of the total variation of the adaptation strategies, in accordance with the KMO rule. (Fig. 3 and Table 7).

Loading the adaptation strategies of the dry season sorghum farmers according to the five (5) main factors gave the results mentioned in Table 8.

Table 5 Results of the loading of the rainy season sorghum farmers' adaptation strategies by factor

Adaptation strategies	Factor 1	Factor 2
Sowing early matured varieties	**−0.570**	0.067
Sowing or transplanting early	−0.444	**−0.511**
Sowing of drought resistant crops varieties	−0.267	**0.513**
Diversification of crops varieties	**−0.712**	0.292
Change of crops or crops varieties	**0.582**	0.475
Ploughing of plots and mounding of plants	**0.862**	−0.217
Temporary or permanent transfer of crops	**0.540**	−0.424
Making of racks	−0.572	**0.456**
Organic or mineral fertilizer input	**0.554**	−0.052
Diversification of income-generating activities	−0.127	**0.697**
Crops' diversification	0.133	**0.163**
Multiplication of weeding	−00.77	**0.690**
Sowing of melted seed holes or dried plants	**03.65**	−0.735

Table 6 Results of the KMO sample adequacy test applied to the dry season sorghum farmers' adaptation strategies

Adaptation strategies	Codes	KMO values
Sowing of early matured varieties	SEVARPRE	0.817
Sowing or transplanting early	SEMISSEC	0.577
Sowing of drought resistant crops varieties	SEMVARES	0.743
Diversification of crops varieties	DIVARCUL	0.782
Change of crops or crops varieties	CHASPVA	0.522
Ploughing of plots and mounding of plants	LABBUTPL	0.801
Temporary or permanent transfer of crops	MUTEDECU	0.540
Making of racks or bunds	CONFCAS	0.771
Crops organic or inorganic fertilizer input	FUMORGA	0.773
Diversification of income-generating activities	DIVACGER	0.731
Crops diversification	DIVERCUL	0.500
Multiplication of weeding	MULTSARC	0.687
Sowing of molten seed holes or dried plants	RESREPIQ	0.621
Late transplanting	SREPITAR	0,444
Deepening piles	APROFPI	0.555
Purchase or request of nurseries	ACHATPE	0.663
Scaling of nurseries over the time	ECHELPEP	0.780
Organic or inorganic fertilization of nurseries	FEORMIPE	0.568
Cleaning and deepening of ponds	CURAMAE	0.753
Water research over long distances	RECHEAGD	0.739
Fertilization of transplanting water	FERTEARE	0.585

Factor 1 groups together the adaptation strategies "sowing of early matured varieties", "sowing of drought-tolerant varieties", "diversification of crop varieties", "ploughing of plots and mounding of plants", "making of racks or bunds", and

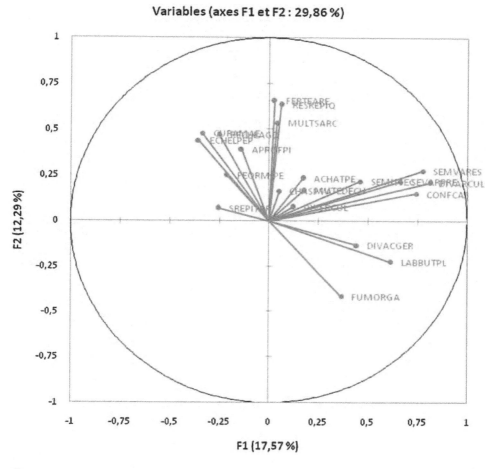

Fig. 3 Percentage of explanations of variables by factors F1 and F2

Table 7 Variability
explained by factor

Factors	Variation	%	Cumulative %
F1	3.690	17.571	17.571
F2	2.581	12.292	29.863
F3	2.132	10.151	40.014
F4	1.561	7.434	47.447
F5	1.295	6.165	**53.612**

"diversification of income-generating activities". This factor brings together all of the farmers' adaptation strategies which aim to avoid or minimize water risks and their impacts, and in fact corresponds to the adaptation strategy by "**a priori eviction or minimization of water risks**" suggested by the first category of typologies.

Factor 2 groups together the adaptation strategies "multiplication of weeding", "sowing of melted seed holes or dried plants", "deepening of piles", "scaling of nurseries over the time", "cleaning and deepening of ponds", "water research over

Table 8 Results of the loading of the dry season sorghum farmers' adaptation strategies by factor

Adaptation strategies	Factor 1	Factor 2	Factor 3	Factor 4	Factor 5
Sowing of early matured varieties	**0.663**	0.219	−0.100	0.194	0.149
Sowing or transplanting early	0.459	0.219	−0.168	−0.163	**0.596**
Sowing of drought resistant crops varieties	**0.777**	0.277	−0.307	0.124	−0.240
Diversification of crops varieties	**0.813**	0.213	−0.246	0.075	−0.160
Change of crops or crops varieties	0.052	0.165	−0.052	**0.487**	−0.248
Ploughing of plots and mounding of plants	**0.613**	−0.224	0.352	−0.070	−0.107
Temporary or permanent transfer of crops	0.176	0.170	0.243	−0.035	**−0.391**
Making of racks or bunds	**0.743**	0.152	−0.349	0.055	−0.171
Crops organic or inorganic fertilizer input	0.367	−0.416	**0.445**	0.029	0.140
Diversification of income-generating activities	**0.443**	−0.133	0.428	0.158	0.387
Crops diversification	0.120	0.077	−0.220	**−0.222**	0.194
Multiplication of weeding	0.039	**0.535**	0.410	−0.154	−0.272
Sowing of melted seed holes or dried plants	0.061	**0.633**	0.464	−0.312	0.019
Deepening of piles	−0.139	**0.392**	−0.287	0.033	0.222
Purchase or request of nurseries	0.172	0.236	0.245	**0.469**	0.297
Scaling of nurseries over the time	−0.357	**0.438**	−0.386	−0.007	0.084
Organic or mineral fertilization of nurseries	−0.215	0.253	0.262	**0.418**	0.245
Cleaning and deepening of ponds	−0.337	**0.477**	−0.359	−0.023	−0.038
Water research over long distances	−0.251	**0.471**	0.049	0.389	0.143
Fertilization of transplanting water	0.026	**0.657**	0.498	−0.401	−0.016

long distances", and "fertilization of transplanting water". This factor brings together all the adaptation strategies aimed at the sustainable management of water resources, and can be called "**Adaptation to water risks by efficient management of water resources**".

Factor 3 contains the "crops organic or inorganic fertilizer input" strategy. It brings together strategies aimed at sustainable soil management, and can be called "**Adaptation to water risks through efficient soil management**".

Factor 4 groups together the strategies "change of crops or crop varieties", "crops diversification", "purchase or request of nurseries", and "organic or mineral fertilization of nurseries". It brings together strategies aimed at the sustainable management of crops, and can be called "**Adaptation to water risks through sustainable management of crops**".

Factor 5 groups together the strategies "sowing or transplanting early" and "temporary or permanent transfer of crops", which aim to avoid water risks, and can be called "**Adaptation by a priori eviction of water risks**".

Analysis of all the five factors reveals that factors 1 and 5 correspond to the farmers' adaptation to climate change by **"eviction or a priori minimization of water risks"**, while factors 2, 3, and 4 correspond to their adaptation by **"confronting water risks"**; and therefore, it could be said that the dry season sorghum farmers' adaptation strategies corresponds very well to the typology proposed by the authors of the first category of typologies, namely, adaptation by "confrontation with the water risks" and adaptation by "a priori eviction or minimization of the water risks."

That said, depending on the results of the characterization of the sorghum farmers' adaptation strategies using PCA and KMO test, their whole adaptation process can be explained through a set of two actions:

1. The agro-pastoral natural resources management by "confrontation with the climate hazards and water risks" or by "eviction of the climate hazards and water risks".
2. The intense spatiotemporal diversification of the practices (agro-pastoral natural resources management, income generating activities).

Finally, it can be said that the characterization of the sorghum farmers' adaptation strategies shows that they are more complex than most authors who have established the typologies thought, because of the spatiotemporal diversification of the practices.

Conclusion

At the end of this chapter, we could draw the following conclusions:

- The poor spatiotemporal distribution of rains and the drought respectively constitute the main climate hazard and the main water risk faced by sorghum farmers in particular, and farmers in general in the semi-arid region of Cameroon.
- The sorghum farmers are highly vulnerable to climate change, and that could be perceived through the coexistence of all the three forms of drought (meteorological, agricultural, hydrological), the permanent food insecurity, the mostly traditional adaptation strategies used and their very low adoption rates, the underuse or absence of efficient adaptation strategies (irrigation, improved crop varieties), and their socioeconomic characteristics (the practice of self-consumption agriculture, the small size of the sown areas, the low quantity of agricultural inputs used, the poor access to agricultural extension and to credits, the multiplication of income-generating activities, and the weak school enrollment rate).
- The characterization of the adaptation strategies used shows that they are more complex than most authors who have established the typologies thought because the whole adaptation process used by sorghum farmers can be explained through a set of two actions: the agro-pastoral natural resources management by "confrontation with the climate hazards and water risks" or by "eviction of the climate hazards and water risks"; and the intense spatiotemporal diversification of the

practices (agro-pastoral natural resources management, income generating activities).

- Insofar, as the farmers are very vulnerable to the climate change, it seems given their poor socioeconomic conditions that a real improvement in their resilience depends absolutely on a real and deep improvement of these socioeconomic conditions.

References

Agossou DSM, Tossou CR, Vissoh VP, Agbossou KE (2012) Perception des perturbations climatiques, savoirs locaux et stratégies d'adaptation des producteurs agricoles béninois. Afr Crop Sci J 20(2):565–588

Arodokoun UA (2011) Socioeconomic impact of the use of NICTs in adaptation strategies to climate change in rural areas: the case of cotton farmers in Center-Benin. Thesis for the Diploma of Agricultural Engineer, University of Abomey-Calavi, Abomey

Batterbury SPJ, Forsyth T (1999) Fighting back: human adaptations in marginal environments. Environment 41(11):25–29

Batterbury SPJ, Mortimore MJ (2013) Adapting to drought in the west African Sahel. In: Boultner S, Palutikov J, Karoly D, Guitart D (eds) Symposium on natural disasters and adaptation to climate change. Cambridge University Press, Cambridge

Blanc E (2012) The impact of climate change on crop yields in sub-Saharan Africa. Am J Clim Chang 1:1–13

Borton J, Nicholds N (1994) Drought and Famine. London, England: FAO

CEDC (2010) Regional master plan for land management and use planning. Technical report. MINPLADAT, Yaoundé

Chédé F (2012) Vulnérabilité et stratégies d'adaptation au changement climatique des paysans du Département des Collines au Bénin: cas de la Commune de Savè. Mémoire de fin d'études pour l'obtention du diplôme de Mastère en Changement climatique et développement durable. Centre Regional AGRHYMET de Niamey, Niger

Diarra A (2008) Adaptation of Sahelian agriculture to climate change: an approach using stochastic modeling. Results of research presented at meetings of the International Institute for Water and Environment Engineering, Ouagadougou

Dingkuhn M (2009) Adaptation des plantes cultivées au changement climatique. In: Caron P (ed) Changement climatique et agriculture: l'environnement et la sécurité alimentaire en jeu. CIRAD, Paris

Fabre C (2010) The adaptation of food farmers in Senegal to climate change: the case of the rural community of Sessène, Thiès region. Master thesis in Geography, University of Montreal, Montreal

FAO (2009) Food production: the determining role of water. Food and Agriculture Organization of the United Nations (FAO), Rome

FAO, NDMC (2008) A review of drought occurrence and monitoring and planning activities in the Near East Region. FAO, Rome

Gnanglé PC, Egah J, Baco MN, Gbemavo CDSJ, Kakai RG, Sokpon N (2012) Perceptions locales du changement climatique et mesures d'adaptation dans la gestion des parcs à karité au Nord-Benin. Int J Biol Chem Sci 6(1):136–149

IPCC (2014) Climate change 2014 Synthesis report: headline statements from the summary for policymakers. Technical report. IPCC, Geneva

Jouve P (2010) Practices and strategies for adapting farmers to climatic hazards in sub-Saharan Africa. Grain de sel 49:15–16

L'Hôte Y (2000) Climatology. In: Seignobos C, Iyébi-Mandjeck O (eds) Atlas of the province of far North Cameroon (Plate 2). IRD, Paris

Leary N, Kulkarmi J, Seipt C (2007) Assessments of impacts and adaptation to climate change: summary of the final report of the AIACC project. Technical report. START, Washington, DC

Mortimore MJ, Adams WM (2000) Farmer adaptation, change and crisis in the Sahel. Glob Environ Chang 11:49–57

Ngigi NS (2009) Climate change adaptation strategies: water resources management options for smallholder farming systems in sub-Saharan Africa. Synthesis report. MDG Centre for East and Southern Africa, Nairobi

Nhemachena C, Hassan R (2007) Micro-level analysis of farmers' adaptation to climate change in southern Africa. IFPRI discussion paper 00714. IFPRI, Washington, DC

OECD (2010) Gestion durable des ressources en eau dans le secteur agricole. OCDE, Paris

Sissoko K, Van Keulen H, Verhagen J, Tekken V, Battaglini A (2010) Agriculture, livelihoods and climate change in the West African Sahel. Reg Environ Change 3:25–37

Yesuf M, Di Falco S, Deressa T, Ringler C, Kohlin G (2008) The impact of climate change and adaptation on food production in low-income countries: evidence from the Nile Basin (Ethiopia). IFPRI discussion paper 00828. IFPRI, Addis Ababa

3

Sustaining a Cleaner Environment by Curbing Down Biomass Energy Consumption

Abubakar Hamid Danlami and Shri Dewi Applanaidu

Contents

A. H. Danlami (✉)
Department of Economics, Faculty of Social Sciences, Bayero University Kano, Kano, Nigeria

S. D. Applanaidu
Department of Economics and Agribusiness, School of Economics, Finance and Banking, College of Business, Universiti Utara Malaysia, Sintok, Malaysia
e-mail: dewi@uum.edu.my

Abstract

Environmental degradation, soil erosion, and desertification are some of the consequences of high rate of traditional biomass fuel use by households in developing countries. The critical issues to raise here are how can these households be encouraged to change their energy consumption behavior? What are the factors that cause the rampant use of biomass fuel in developing countries? How and to what extent can these factors be manipulated so that households in developing countries are encouraged to adopt clean energy fuel an alternative to the most widely used biomass fuel? Therefore, this chapter tries to find answer to the above questions raised, by carrying out an in depth analysis of households' use of biomass fuel in developing countries using Bauchi State, Nigeria, as the case study. Cluster area sampling technique was utilized to generate the various responses, where a total number of 539 respondents were analyzed. The study estimated ordered logit model to analyze the factors that influence the movement of households along the energy ladder from nonclean energy to the cleaner energy. Furthermore, Ordinary Least Squares (OLS) model was estimated to analyze the impacts of socio-economic, residential, and environmental factors on biomass energy consumption. It was found that age of the household head and his level of education, income, living in urban areas, home ownership, and hours of electricity supply have positive and significant impact on household energy switching from traditional biomass energy use to the cleaner energy. Therefore, policies that will enhance household income and the increase in the availability of cheap cleaner energy will encourage households switching to cleaner energy sources thereby reducing the level of environmental pollution in the study area.

Keywords

Clean-energy · Households · OLS, Ordered-logit · Traditional-biomass

Introduction

Climate change which is the variation in climate overtime and its impacts on the environment and socio-economic systems now constitute the most important environmental problem facing mankind. According to UNDP (2013), climate change can affect both the human and the entire natural systems which pose a threat to human development and survival. Among the most important factors that caused climate change and environmental pollution is the wide spread use of biomass energy. Biomass fuels, which constitute animal dung, crop residues, fuel-wood, and charcoal, are among some of the most widely used fuels for cooking and heating, particularly in developing countries (Yamamoto et al. 2009). Currently, more than two billion people rely on biomass fuel globally to satisfy their basic energy needs. That is why this type of fuel accounts for about 20% of energy supply for the whole world. IEA (2011) argued that if proper measures were not taken, the number of biomass fuel users globally may increase to about 2.7 billion people by the year 2030. A large number of biomass energy users, especially in traditional ways, are found mostly in Asia and Africa. For instance, the IEA (2011) reported that biomass fuel accounts for more than 50% of Africa's energy utilization. In Nigeria, the wide use of biomass fuel at household level is more than 70%. Coming down to the state level, the rate of biomass fuel use by households is about more than 90% in Bauchi State, Nigeria (NBS 2012).

However, such wider use of biomass fuel is unwholesome for human living and their environment. For instance, when combining CO_2 emissions and other GHG in a single index, biomass fuel scores much higher than other fossil fuels like LPG and kerosene. Moreover, there is higher correlation between the use of biomass fuel and indoor air pollution (Desalu et al. 2012; Risseeuw 2012). Such indoor air pollution causes about 1.5 to 2 million losses of lives worldwide yearly. Environmental degradation, soil erosion, and desertification are some of the consequences of high rate of biomass fuel use in developing countries. In fact, Nigeria has been facing an annual average rate of deforestation over a previous decade due to high rate of biomass fuel use. Available data has shown that the nation's 15 million hectares of forest and woodland reserves could be depleted within the next 50 (Nnaji et al. 2012). In Bauchi State, the rampant use of biomass energy is so great to the extent of more than 600 kg per household monthly (Danlami et al. 2017). This has posed negative impacts on the inhabitant of the State such as the systematic destruction of the State's forest reserves and woodlands, soil erosion, and desertification whereby the State losses not less than 1 km of land yearly due to desertification as a result of high rate of felling trees (Danlami et al. 2017).

That is why the defunct United Nations Millennium Project recommended halving the number of households that depend on traditional biomass fuel by the year 2015, the target that was not complied by most of the participating countries including Nigeria (Naibbi and Healey 2013). The critical issues to raise here are that how can these households be encouraged to change their energy consumption behavior? what are the factors that cause the rampant use of biomass fuel in developing countries? how and to what extent can these factors be manipulated so that households in developing countries are encouraged to adopt clean energy fuel an alternative to the most widely used traditional biomass fuel? This is because clean fuel has greater capacity to do useful work. The use of clean fuel is an imperative to improve the standard of living of the households that heavily rely on biomass fuel (Lee 2013). Therefore, in line with the above issues raised, this chapter tries to find answer to the above questions raised by carrying out an in depth analysis of households' use of biomass fuel in developing countries using Bauchi State, Nigeria, as the case study. The remaining part of the chapter consists sections as follows. Section "Literature Review" constitutes review of related literature. Section "Theoretical Frameworks" is theoretical and conceptual frameworks of the study. Section "Methodology" highlights the methodology adopted. Section "Analysis" the discussion of findings. Section "Conclusions" highlights chapter conclusion. The last section constitutes the policy implications and recommendations.

Literature Review

Studies that analyzed energy utilization can be classified into two categories. The first group consists of those studies (Danlami et al. 2018a, 2019a) that analyze aggregate energy consumption using time series data. Most of these studies concluded that energy consumption is highly correlated with environmental degradation. However, these conclusions have limited practical applicability at a microlevel, because the energy consumption behavior of households is heterogeneous, which is usually ignored in studies that utilized time series data. Moreover, the second category of studies are those that analyzed household energy consumption using microdata approach (Lee 2013; Mensah and Adu 2013; Ozcan et al. 2013; Abdurrazak et al. 2012; Couture et al. 2012; Laureti and Secondi 2012; Onoja 2012; Oyekale et al. 2012; Song et al. 2012; Ganchimeg and Havrland 2011; Jingchao and Kotani 2011; Jumbe and Angelsen 2010; Osiolo 2010; Suliman 2010; Danlami and Islam 2020; Danlami et al. 2017; Nlom and Karimov 2014; Ogwumike et al. 2014). These studies arrived at different conclusions based on the socio-economic, demographic, home, and environmental characteristics of the households under consideration.

The composition and type of socio-demographic factors of households determine their fuel switching and consumption behavior. For instance, Laureti and Secondi (2012) indicated that households which comprise of couples with children tend to adopt less of oil and electricity and more of coal-wood when compared with a household of a single person. This is contrary to the findings by Danlami et al. (2017) which concluded that the household that is headed by a married individual has higher odd of adopting clean fuel than otherwise. Whereas some previous studies (Osiolo 2010; Jumbe and Angelsen 2010; Nlom and Karimov 2014) reported no significant relationship between the gender of the household head and its energy consumption behavior, a study by Mensah and Adu (2013) found that there is a significant negative relationship between the household head being male and the adoption of clean energy. Meanwhile, age of the household head was found to have a negative impact on the adoption of clean energy (Suliman 2010; Mensah and Adu 2013; Nlom and Karimov 2014). Households adopt less clean energy source when the head is older. Additionally, the household head level of education was found to exact a positive impact on the adoption of clean energy. The higher the level of education of the household head, the lower the probability of adopting nonclean energy (Eakins 2013; Mensah and Adu 2013; Ozcan et al. 2013). The number of a household's members (i.e., household size) affects the household's energy switching decision, the larger the size of a household, the lesser the adoption of clean energy. This assertion is supported by previous studies (Ozcan et al. 2013; Mensah and Adu 2013; Suliman 2010; Heltberg 2005).

The factors that measure the economic status of the household influence the households' fuel consumption decision. For instance, studies have established that there is a positive relationship between income and adoption of clean energy (Danlami et al. 2017; Mensah and Adu 2013; Ozcan et al. 2013; Couture et al. 2012). Poorer households especially in developing countries tend to adopt biomass fuels like firewood, plant residues, and animal dung. Furthermore, number of energy consuming appliances increases the quantity of energy consumption by households (Danlami 2017a; Eakins 2013). The higher the number of energy consuming appliances at home, the lesser the odd of adopting biomass source of energy (Danlami 2017b). Moreover, energy price has a negative relationship with energy consumption. When the price of a particular energy source is high, households switch to other alternative fuel available. This is in line with law of demand and also has been established by previous studies (Danlami 2017a; Nlom and Karimov 2014; Lee 2013).

Furthermore, the characteristics of the building in which the households live also affect their energy choice behavior. Factors such as size of the building, number of rooms in the home, share of dwelling and dwelling ownership have been established by previous studies to influence the manner of household energy consumption behavior (Danlami et al. 2017; Eakins 2013; Mensah and Adu 2013; Tchereni 2013). Lastly, environmental factors such as location of home, the extent of electricity supply, the main source of cooking, and lighting fuel in the area were found to influence the manner of household energy consumption. Households that live in the rural areas or the area whereby there is a wide spread use of biomass fuel tend to adopt biomass fuel as their main source of energy (Danlami 2017a, b; Ozcan et al. 2013).

Based on the above-reviewed literature, it can be seen that there exist inconsistences as per the findings and conclusions of previous studies on the factors influencing household energy consumption and switching behavior, from one place to another, due to differences in environmental factors, cultural factors, socio-economic settings, as well as differences in the average level of development among different regions. Therefore, additional study on household energy choice and consumption in a specific area is an addition to the existing literature as argued by the previous studies (Danlami et al. 2015, 2017).

Theoretical Frameworks

Households mostly use energy for indirect satisfaction mainly to produce another commodities or services (modified from Danlami et al. 2016). Households utilized energy from different sources for the purpose of maximizing satisfaction. This optimal level satisfaction is usually attained at the equilibrium point of particular energy consumption. For instance, equation (1) indicates a given utility function of energy consumption:

$$U = f\left(\in_c G_s \mathcal{L}_f \mathcal{B}_b\right) \tag{1}$$

Subject to household budget constraint as in equation (2):

$$Y = P_c \in_c + P_s G_s + P_f \mathcal{L}_f + P_b \mathcal{B}_b \tag{2}$$

where U = utility, $\in_c$ = electric energy, G_s = gas energy, L_f = liquid energy (i.e., fuel), B_b = biomass energy, Y = income of household and P = price of the relevant energy.

To find the maximum point of household energy utilization, we form Langarangian multiplier function as in equation (3):

$$L = f(\in_c G_s \; L_f \; B_f) + \lambda(Y - P_c \in_c + P_s G_s + P_f L_f + P_b B_b = 0) \qquad (3)$$

Using equation (3), we can analyze the maximum point of utility for:

(i) Household that use only one of these energy sources
(ii) Households that use all of these energy sources

Equilibrium Level of Utility for Households That Use Only One of These Energy Sources

Assuming the households use only electricity as its sole source of energy, the utility maximization point will be

$$\frac{\partial L}{\partial \in_c} = f'_c - \lambda P_c = 0 \qquad (4)$$

$$f'_c = \lambda P_c \qquad (5)$$

Since the household utilized only single source of energy, $\lambda = 1$

$$f'_c = P_c \qquad (6)$$

Equation (6) indicates the point of utility maximization from using electric source of energy where the marginal utility obtained from consuming extra unit of electricity is equal to the price of that additional unit of electricity. Any increase in the consumption of electricity above the equilibrium level implies decrease in the total utility, while consumption of electricity below the equilibrium level implies that the total utility of electricity is not maximized because additional unit of electricity consumed will lead to increase in the total utility, until the above equilibrium point is reached.

Equilibrium Point for Households That Use Gas as Their Only Source of Energy

In this case, we conduct the partial derivation of equation (3) with respect to gas. This is indicated in equation (7):

$$\frac{\partial L}{\partial G_s} = f'_s - \lambda P_s = 0 \qquad (7)$$

$$f'_s = \lambda P_s \qquad (8)$$

By definition, $\lambda = 1$, therefore, the utility maximization point will be

$$f'_s = P_s \qquad (9)$$

that is, the point where the additional satisfaction obtained from using extra amount of gas is equal to the price of that additional unit of gas.

Derivation of Equilibrium Point for Households That Use Liquid Fuel as the Only Source of Energy

Here, we find the partial derivative of equation (3) with respect to the liquid fuel as in equation (10):

$$\frac{\partial L}{\partial L_f} = f'_f - \lambda P_f = 0 \qquad (10)$$

$$f'_f = \lambda P_f \qquad (11)$$

Since $\lambda = 1$ (for households that use only one source of energy)

$$f'_f = P_f \qquad (12)$$

Utility Maximization Point for Households That Use Only Biomass Energy

The partial derivative of equation (3) with respect to biomass energy is given by:

$$\frac{\partial L}{\partial \mathcal{B}_b} = f'_b - \lambda P_b = 0 \qquad (13)$$

$$f'_b = \lambda P_b \qquad (14)$$

Since $\lambda = 1$

$$f'_b = P_b \qquad (15)$$

Equation (15) indicates the utility maximization point for household that utilizes only biomass energy. This is the point where the additional satisfaction obtained from using an additional bundle of biomass energy is equal to the price of that additional bundle.

Utility Maximization of Households That Use All the Four Source of Energy Together

In this situation, the utility of using energy is maximized, by consuming the energy up to the level where the ratio of extra satisfaction from using the additional amount of energy to their prices is equal. Taking back the earlier Langarangian multiplier utility function and the constraints for energy use

$$L = f\left(\in_c G_s \, \mathcal{L}_f \, \mathcal{B}_b\right) + \lambda\left(Y - P_c \in_c + P_s G_s + P_f \mathcal{L}_f + P_b \mathcal{B}_b = 0\right) \qquad (16)$$

The partial derivatives with respect to each of the energy source are:

$$\frac{\partial L}{\partial \in_c} = f'_c - \lambda P_c = 0 \qquad (17)$$

$$\frac{\partial L}{\partial G_s} = f'_s - \lambda P_s = 0 \qquad (18)$$

$$\frac{\partial L}{\partial \mathcal{L}_f} = f'_f - \lambda P_f = 0 \qquad (19)$$

$$\frac{\partial L}{\partial \mathcal{B}_b} = f'_b - \lambda P_b = 0 \qquad (20)$$

$$\lambda = \frac{f'_c}{P_c} = \frac{f'_s}{P_s} = \frac{f'_f}{P_f} = \frac{f'_b}{P_b} \qquad (21)$$

That is the utility maximization point for households that use all the four source of energy is for them to consume at the point where the ratio of the extra satisfaction from using additional unit from each of the energy source to their prices are equal.

Methodology

Following Danlami et al. (2019b), the total sample size was determined based on Dillman (2011). A total of 750 questionnaires were distributed based on cluster area sampling method. Finally about 548 filled questionnaires were returned back (which is more than 70% of the total number of the issued questionnaires) out of which 9 questionnaires were discarded. Multistage cluster sampling was utilized as the sampling technique.

Model Specification

Ordered Logit Models
In order to satisfy the first objective of this chapter which is to assess the determinants of household fuel switching up the ladder from traditional biomass energy to the cleaner source of energy, ordered logit model was employed. Since household fuel switching consists of movement up the energy ladder in a hierarchical order which is the basis for ordered models (Kofarmata 2016). Therefore, due to the ordinal nature of the dependent variable, it is stated as movement in fuel the household switching from traditional biomass energy, transitional energy (kerosene), and the cleaner energy (gas and electricity sources of energy). Thus, the model can be stated as in equation (22):

$$y_i = \beta X_i + \varepsilon_i \tag{22}$$

where y_i is the observed and exact dependent variable (categories of fuel switching in hierarchical order); coded as 0, 1, ...n, X_i is the vector of the independent variables. β is the vector of parameters to be estimated and ε_i is the random variable for the ordered logit model.

If the score on the observed variable say y_i is 0, means that the household uses traditional biomass energy. However, if the household adopts the transitional fuel (such as kerosene), then $y_i = 1$; and if the household adopts cleaner source of energy (electricity/gas) then $y_i = 2$. Then the estimated empirical model is written as:

$$\begin{aligned}
Y_i = {} & \alpha_0 + \beta_1 \text{GEN}_i + \beta_2 \text{AGE}_i + \beta_3 \text{EDU}_i + \beta_4 \text{HHS}_i + \beta_5 \text{INC}_i + \beta_6 \text{LOC}_i \\
& + \beta_7 \text{NRM}_i + \beta_8 \text{DSH}_i + \beta_9 \text{HRSE}_i + \beta_{10} \text{PFW}_i + \beta_{11} \text{NCF}_i \\
& + \beta_{12} \text{HAPP}_i + \beta_{13} \text{HOWN}_i + \varepsilon_i
\end{aligned} \tag{23}$$

Y_i = The dependent ordered variables summarized as: Traditional biomass ($y_i = 0$), Transitional energy ($y_i = 1$) and Cleaner energy ($y_i = 2$)
GEN_i = Gender of the head of household
AGE_i = Age of the head of household
EDU_i = Level of education of the head of household
HHS_i = Size of the household
INC_i = Monthly income of the head of household
LOC_i = Home location of the household
NRM_i = Number of rooms in the home
DSH_i = Size of the dwelling of the household
HRSE_i = Hours of electricity supply

PFW_i = Unit price of firewood per bundle
NCF_i = Similarity with the neighbor's main cooking fuel source
$HAPP_i$ = Home appliances
$HOWN_i$ = Home ownership

The OLS Model

Another objective of this study is to assess the determinants of household for biomass energy in Bauchi State.

Following Danlami (2014) and Lee (2013), the implicit form of the relationship between households' consumption of a particular energy and its determinants can be expressed as:

$$Y_i = \beta_0 + \sum_{i=0}^{k} \beta_i X_i \qquad (24)$$

where Y_i is household i's consumption of biomass energy.

The estimated empirical OLS model for households' biomass energy consumption is expressed as:

$$ln\,FWD_i = \alpha_0 + \beta_1 GEN_i + \beta_2 AGE_i + \beta_3 MST_i + \beta_4 EDU_i + \beta_5 HHS_i$$
$$+ \beta_6 INC_i + \beta_7 PFW_i + \beta_8 PKR_i + \beta_9\,ln\,HAPP_i + \varepsilon_i \qquad (25)$$

where
FWD_i = Quantity of firewood bundle consume monthly.
GEN_i = Gender of the head of household
AGE_i = Age of the household head
MST_i = Marital status of the head of household
EDU_i = Level of education of the head of household
HHS_i = Size of the household
Inc_i = Monthly income of the head of household
PFW_i = Unit price of firewood per bundle
PKR_i = Price of kerosene per liter
$HAPP_i$ = Number of home appliances own by household

Analysis

This study mainly analyzes two issues: household energy switching and the extent of household traditional biomass energy use. Table 1 indicates the estimated ordered logit model analyzing the determinants of household energy switching. Furthermore, Table 2 exhibits the estimated OLS model for the determinants of household

Table 1 Estimated coefficients of energy switching (ordered logit model)

Variables	Coefficients
Gender	0.7079
	(0.5350)
Age	0.0233*
	(0.0126)
Education	0.0869**
	(0.0416)
Household size	−0.1148***
	(0.0376)
Lnincome	0.0106***
	(0.0040)
Location	0.6118*
	(0.3479)
Number of rooms	−0.0512
	(0.0388)
Dwshare	−0.0075
	(0.2634)
Hours of electricity supply	0.0149***
	(0.0041)
Firewood price	0.0024
	(0.0038)
Ncfuel	−1.2996***
	(0.2905)
Home appliances	−0.0023
	(0.0103)
Home ownership	0.6069**
	(0.2879)
	(1.124)
Observations	444
Pseudo R^2	0.21

$\chi^2 (26) = 72.56$
Prob $> \chi^2 = 0.0000$
Note: Robust standard errors in parentheses *** $p < 0.01$, ** $p < 0.05$, * $p < 0.1$

traditional biomass consumption. The analysis and discussions of the estimated models are carried out in the following sections:

Determinants of Household Energy Switching

The first objective of this study is to assess the factors that determine the household movement along the energy ladder from traditional biomass energy to the available cleanest energy (electricity) in Bauchi State. To achieve this objective, ordered logit model was estimated and the result of the estimation is indicated in Table 1.

Table 2 Determinants of household traditional biomass energy use

Variables	Coefficients	Standard error
Gender	0.1190	(0.0834)
Age	0.0012	(0.0022)
Marital status	−0.1414**	(0.0598)
Education	−0.0078*	(0.0046)
Household size	0.0168***	(0.0041)
lnIncome	0.0382	(0.0420)
Price of firewood	−0.0030**	(0.0013)
Price of Kerosene	−0.0034***	(0.0010)
lnhappls	−0.262	(0.0496)
Constant	3.9106***	(0.241)
Observations	270	
R^2	0.13	

Ramsey RESET Test (Specification test)
$F(3, 244) = 5.63$
Prob $> F = 0.000$
Note: Robust standard errors in parentheses*** $p < 0.01$, ** $p < 0.05$, * $p < 0.1$

Based on the probability value of the Chi sq (X^2), in Table 1, the estimated ordered logit model is jointly significant at 1%, thereby implying the validity of the estimated ordered model. The result indicates that the coefficient of variable Age is statistically significant at 10% level. The result indicates that there is a positive relationship between the age of the household head and the household energy switching. A 1 year increase in the age of the household will lead to a 0.023 log odd of household being in higher level of cleaner energy. This is in line with a priori expectation and conforms to the findings of Danlami (2017c). In the same vein, the coefficient of education was found to have a positive relationship with the household energy switching. The coefficient of the variable was found to be statistically significant at 5% level. The estimated result shows that an increase in the level of education of the household head increases the log odd of household switching to cleaner energy by about 0.087 units. This is in line with a priori expectation because when the household head is more educated, he will have more awareness about the negative effects of using traditional biomass energy. This result supports the findings from Danlami et al. (2018b).

Contrarily, the coefficient of household size was found to have a negative relationship with the household energy switching. This coefficient was found to be statistically significant at 1% level. Based on the estimated result, a one unit increase in the number of household size decreases the log odd of switching to cleaner energy by about 0.115 units. This is in line with the findings of Danlami et al. (2019c). The coefficient of household income was found to be statistically significant at 1% level. The estimated result indicates that a 1% rise in the household income leads to increase in the log odd of household switching to the cleaner energy by about 0.011 units. This is in line with a priori expectation because as the household income increases, the affordability of the household to substitute traditional biomass energy with a cleaner energy increases. This supports the findings of Danlami (2017a). Similarly, the coefficient of household location was found to have a positive relationship with the household energy switching and it was found to be statistically significant at 10% level. Based on the result of the estimated ordered logit model, households that reside in urban areas of Bauchi State have higher log odd of switching to cleaner energy by about 0.612 units compared to their rural counter parts. This is in line with a priori expectations that the households living in urban areas adopt cleaner energy than the households living in the rural areas mainly due to economic, social, and educational factors. The exact fact is that cleaner sources of energy are more available in urban than in rural areas, which is in line with the findings of Danlami et al. (2018c).

The coefficient of hours of electricity supply was found to be statistically significant at 1% level and it was found to have a positive relationship with household energy switching. Based on the estimated model, 1 h increase in electricity supply increases the households' log odd of switching to cleaner energy by about 0.015 units. This conforms to a priori expectation that the tendency of households to move up the cleaner energy ladder increases when the electricity supply becomes more available and reliable. This supports the findings of Danlami et al. (2018b). Finally, the coefficient of home ownership was found to be statistically significant at 5% level. Based on the estimated ordered logit model, there is a positive relationship between the coefficient of home ownership and switching to cleaner energy in Bauchi State. The households that live in their self-owned home have higher log odd of switching to cleaner energy by about 0.607 units than otherwise. This is in line with the findings of previous studies (Danlami et al. 2018b)

Determinants of Household Traditional Energy Use

Another objective of this study is to assess the factors that influence the quantity of traditional biomass energy use in Bauchi State, Nigeria. The result of the estimated model is in Table 2.

Based on Table 2, the estimated result shows that overall, the model is statistically significant at 1% level with an estimated F-value = 5.63 and the corresponding probability value Prob(F) = 0.000. The result in Table 2 has shown that the estimated coefficient of marital status is statistically significant at 5% level. On average, the households that are headed by a married person consume less traditional biomass energy by about 14% lower compared to the households that are headed by a nonmarried person. This does not conform to a priori expectation because the expectation is that when the head of a household is married, it means more number of household members which necessitates the use of more traditional biomass fuel such as firewood. However, this may be because the married household head in some cases signifies that he is at least more economically stronger to buy cleaner fuel than firewood. Based on the culture of the people in the study area, a person usually married when economically can afford the marriage responsibilities of which the purchase of cooking fuel is among. This finding supports the findings of other previous studies (Danlami 2019).

Moreover, the coefficient of level of education of the household was also found to be negative and statistically significant at 10% level. Based on the estimated OLS model, an additional year in the level of education reduces the household's use of traditional biomass energy by about 0.78%. This conforms to a priori expectation that as more educated is the household head, the more he has health consciousness and also the more he knows the risk of using traditional biomass energy thereby minimizing the use of such energy. This is in line with the findings of Danlami (2017a) and Lee (2013). On the other hand, the coefficient of household size was found to be statistically significant at 1% and positively related to household use of traditional biomass energy. Based on the result shown in Table 2, increase in the number of household by one individual increases the household's use of traditional biomass energy by about 1.68%. This conforms to a priori expectation and is in line with the findings of previous studies (Danlami 2017a).

The result also indicates that there is a negative significant relationship between traditional biomass energy use and its price. A one Naira increase in the price of firewood decreases the rate of household traditional biomass energy use by about 0.3% all things being equal. This is tally with a priori expectation because as the price of the traditional biomass energy increases, the household will switch to the use of available cheaper and cleaner energy. Similarly, when the price of a commodity rises, the purchasing power of buyers decreases, leaving the consumer with the ability to buy less of that commodity. This finding is in line with traditional law od demand which says that the higher the price, the lower the quantity demanded and also supports the findings of. Lastly, the result shows that price of kerosene has a negative impact on traditional biomass energy.

Conclusions

This study analyzes household energy switching along the energy ladder using ordered logit model. Also, the study uses OLS regression model to analyze the determinants of household traditional biomass energy use. The age of the household head and his level of education, income, living in urban areas, home ownership, and hours of electricity supply have positive significant impact on household energy switching from traditional biomass energy use to the cleaner energy. On the other hand, household size was found to have a negative relationship with household energy switching. Furthermore, the estimated OLS model indicates that household size has a positive and significant impact on traditional biomass energy use, the higher the household size, the high the quantity of traditional biomass energy consumption all things being equal. Marital status, household head level of education, and the price of the traditional biomass energy have negative significant impact on household use of traditional biomass energy.

Policy Recommendations

Having conducted empirical investigation of household energy switching and traditional energy consumption in Bauchi State, Nigeria, the following recommendations were offered based on the study findings, in order to encourage households to switch to cleaner energy in the study area. Since increase in income was found to have significant impact in encouraging households' energy switching up to cleaner energy, policies and programs aimed at raising income earnings of individuals should be embarked upon to discourage the adoption and use of traditional biomass energy. Income can be increased via employment generation, wealth creation, increase in government expenditure, empowering small and medium scale industries, and skills acquisition and development programs.

The study finds that households that live in urban areas have higher probability and odd of switching to cleaner energy. In line with this finding, government should try to make cleaner energy available and affordable to rural dwellers as is in the urban areas. All the facilities that will ensure the availability of cheap cleaner energy in rural areas of the State should be established in order to encourage households to switch to cleaner energy in rural areas of the State.

The findings revealed that the level of formal education attainment by the household head has significant influence on switching to cleaner energy, the higher the level of education of the household head, the higher the odd of switching to the cleaner energy. Therefore, government should embarked upon policies to encourage higher education attainment of people leaving in the study area, especially rural areas whereby there are a large number of illiterate people. High rate of school enrolment can be increased via policies like free universal basic education programs, higher education enrolment at a subsidized rate, construction of more schools near to the people especially in rural areas, provision of more scholarships at higher levels, employing adequate number of teachers to meet the growing number of pupils, and increase in expenditure on educational facilities. The curriculum of the educational system should emphasize on the danger of high rate of environmental pollution and contamination especially in rural areas whereby the rate of awareness is very low.

Lastly, the study has found that adequate supply of electricity has significant impact on household switching to cleaner energy use. Therefore, provision of cheap and adequate electricity supply to households will encourage many households to use electricity as their main source of cooking and lighting, thereby reducing the rate of traditional biomass energy use.

References

Abdurrazak NTA, Medayese SO, Martins VI, Idowu OO, Adeleye BM, Bello LO (2012) An appraisal of household domestic energy consumption in Minna, Nigeria. J Environ Sci Toxicol Food Technol 2:16–24

Couture S, Garcia S, Reynaud A (2012) Household energy choices and fuelwood consumption: an econometric approach using French data. Energy Econ 34:1972–1981

Danlami AH (2014) Examination of determinants of demand for Fertiliser in Tofa local government area, Kano State. Niger J Manag Technol Dev 5(2):1–14

Danlami AH (2017a) An analysis of household energy choice and consumption in Bauchi State, Nigeria [thesis]. School of Economics Finance and Banking: Universiti Utara Malaysia

Danlami AH (2017b) Determinants of household electricity consumption in Bauchi State, Nigeria. Hyper Econ J 5(1):16–28

Danlami AH (2017c) An intensity of household kerosene use in Bauchi State, Nigeria: a tobit analysi. Niger J Manag Technol Dev 8(2):1–13

Danlami AH (2019) Assessment of factors influencing firewood consumption in Bauchi State, Nigeria. J Sustain Sci Manag 14(1):99–109

Danlami AH, Islam R (2020) Explorative analysis of household energy consumption in Bauchi State, Nigeria. In: Editor A (ed) Energy efficiency and sustainable lighting – a bet for the future. IntechOpen, London

Danlami AH, Islam R, Applanaidu SD (2015) An analysis of the determinants of household energy choice: a search for conceptual framework. Int J Energy Econ Policy 5(1):197–205

Danlami AH, Islam R, Applanaidu SD, Tsauni AM (2016) An empirical analysis of fertiliser use intensity in rural Sub-Saharan Africa: evidence from Tofa local government area, Kano State, Nigeria. Int J Soc Econ 43(12):1400–1419

Danlami AH, Applanaidu SD, Islam R (2017) From biomass cooking fuel source to modern alternative for Bauchi State households: a preliminary analysis. Biofuels 8(3):323–331

Danlami AH, Applanaidu SD, Islam R (2018a) Movement towards a low carbon emitted environment: a test of some factors in Malaysia. Environ Dev Sustain 20(3):1085–1102

Danlami AH, Applanaidu SD, Islam R (2018b) An analysis of household cooking fuel choice: a case of Bauchi State, Nigeria. Int J Energy Sector Manag 12(2):265–283

Danlami AH, Applanaidu SD, Islam R (2018c) Axiom of the relative income hypothesis and household energy choice and consumption in developing areas: empirical evidence using Verme model. Kasetsart J Soc Sci 39:422–431

Danlami AH, Aliyu S, Danmaraya IA (2019a) Energy production, carbon emissions and economic growth in lower-middle income countries. Int J Soc Econ 46(1):97–115

Danlami AH, Applanaidu SD, Islam R (2019b) A micro level analysis of the adoption and efficiency of modern farm inputs use in rural areas of Kano State, Nigeria. Agric Res 8(3):392–402

Danlami AH, Applanaidu SD, Islam R (2019c) Movement towards the adoption of non-traditional household lighting fuel energy in developing areas. Biofuels 10(5):623–633

Desalu O, Ojo O, Aritibi E, Kolawale F, Idowu A (2012) A community survey of the pattern and determinants of household sources of energy for cooking in rural and urban South Western, Nigeria. Pan Afr Med Res J 2:1–12

Dillman DA (2011) Mail and internet surveys: the tailored design method 2007 update with new internet, visual, and mixed-mode guide. Hoboken, Wiley

Eakins J (2013) An analysis of the determinants of household energy expenditures: empirical evidence from the Irish household budget survey. PhD thesis, University of Surrey

Ganchimeg G, Havrland B (2011) Economic analysis of household energy consumption: the case of herders in Mongolia. Agric Trop Subtrop 44:197–203

Heltberg R (2005) Factors determining household fuel choice in Guatemala. Environ Dev Econ 10: 337–361

International Energy Agency (IEA) (2011) Energy balances of non-OECD countries. IEA, Paris

Jingchao Z, Kotani K (2011) The determinants of household energy demand in rural Beijing: can environmentally friendly technologies be effective? Energy Econ 34:381–388

Jumbe BLC, Angelsen A (2010) Modeling choice of fuelwood source among rural households in Malawi: a multinomial probit analysis. Energy Econ 33:732–738

Kofarmata YI (2016) An economic analysis of participation in credit market and credit rationing among farmers in Kano State, Nigeria [thesis]. School of Economics Finance and Banking, Universiti Utara Malaysia

Laureti T, Secondi L (2012) Determinants of households' space heating type and expenditures in Italy. Int J Environ Res 6(4):1025–1038

Lee LY (2013) Household energy mix in Uganda. Energy Econ 39:252–261

Mensah T, Adu G (2013) An empirical analysis of household energy choice in Ghana. Uppsala working paper series no 6

Naibbi AI, Healey RG (2013) Northern Nigeria's dependence on fuelwood: insights from nation-wide cooking fuel distribution data. Int J Humanit Soc Sci 3:160–173

NBS (2012) Annual abstract: federal republic of Nigeria. Retrieved February, 2016 from: http://www.nigerianstat.gov.ng

Nlom JH, Karimov AA (2014) Modeling fuel choice among households in Northern Cameroon. WIDER working paper series, 2014/038.45

Nnaji C, Ukwueze E, Chukwu J (2012) Determinants of household energy choices for cooking in rural areas: evidence from Enugu State, Nigeria. Cont J Soc Sci 5(2):1–11

Ogwumike FO, Ozughalu UM, Abiona GA (2014) Household energy use and determinants: evidence from Nigeria. Int J Energy Econ Policy 4(2):248–262

Onoja AO (2012) Econometric analysis of factors influencing fuel wood demand in rural and Peri-urban farm households of Kogi State. J Sustain Dev 8(1):115–127

Osiolo HH (2010) Enhancing household fuel choice and substitution in Kenya. Kenya Institute for Public Policy Research, Nairobi

Oyekale A, Dare A, Olugbire O (2012) Assessment of rural households' cooking energy choice during kerosene subsidy in Nigeria: a case study of Oluyole local government area of Oyo State, Nigeria. Afr J Agric Res 7:5405–5411

Özcan KM, Gülay E, Üçdoğruk S (2013) Economic and demographic determinants of household energy use in Turkey. Energy Policy 60:550–557

Risseeuw N (2012) Household energy in Mozambique: a study on the socioeconomic and cultural determinants of stove and fuel transitions. Unpublished Master's thesis, Vrieje University, Amsterdam

Song N, Arguilar FX, Shifley SR, Goerndt ME (2012) Factors affecting wood energy consumption by U.S. households. Energy Econ 34:389–397

Suliman MK (2010) Factors affecting the choice of households' primary cooking fuel in Sudan. Research report presented to the Economic Research Forum, Cairo

Tchereni BH (2013) A microeconomic analysis of energy choice behaviour in South Lunzu Township, Malawi. Mediterr J Soc Sci 4(6):569–578

UNDP (2013) Human development report: Nigeria. Retrieved 22 November, 2014 from http://resourcedat.com/wp-content/uploads/2013/05/Nigeria-HDI-value

Yamamoto S, Sie A, Sauerborn R (2009) Cooking fuels and the push for cleaner alternatives: a case study from Burkina Faso. Glob Health Action 2:156–164

4

Climate Change and Variability on Food Security of Rural Household: Central Highlands, Ethiopia

Argaw Tesfaye and Arragaw Alemayehu

Contents

Abstract

This chapter analyzes the impact of climate change and variability on food security of rural households in the central highlands of Ethiopia taking Basona Werana district as a case study site. Data were obtained from 123 households selected using simple random sampling from three agro ecological zones. Key informant interviews and focus group discussion (FDG) were used to supplement the data obtained from household survey. The monthly rainfall and temperature data are for 56 points of 10×10 km grids reconstructed from weather stations

A. Tesfaye (✉)
Department of Geography and Environmental Studies, Mekdela Amba University, Mekane Selam, Ethiopia

A. Alemayehu
Department of Geography and Environmental Studies, Debre Berhan University, Debre Berhan, Ethiopia

and meteorological satellite observations, which cover the period between 1983 and 2016. Standardized rainfall anomaly (SRA), linear regression (LR), and coefficient of variation (CV) are used to examine inter-annual and intra-annual variability of rainfall. Annual and seasonal rainfalls show decreasing trends over the period of observation. The decreasing trends in annual and March–May (Belg) rainfall totals exhibit statically significant decreasing trends at $p = 0.05$ level. Kiremt (June–September) shows statically significant decreasing trends at $p = 0.1$ level. Mean annual maximum and minimum temperatures show statically significant increasing trends at $p = 0.05$ level. More than 80% of households perceived that the climate is changing and their livelihoods (crop and livestock production) are impacted. The district belongs to one of the most vulnerable areas to climate change and variability in the country where large proportions of households (62%) are under different food insecurity classes. Results suggest that local level investigations are useful in developing context-specific climate change adaptation.

Keywords

Climate variability · Rainfall and temperature trends · Food security · Ethiopia

Introduction

Africa is already affected by climatic extremes such as floods, droughts, and windstorms, which are aggravated by climate change and variability. Ethiopia is one of the African countries which is most vulnerable to drought and floods (Belay et al. 2017). Climate change in the form of rising temperature, variable rainfall, droughts, and flooding affects agricultural production and threatens food security in low-income and agriculture-based economies (Mongi et al. 2010; Mesike and Esekhade 2014). Agriculture in Ethiopia is a major source of food and contributing sector to food security. It plays a key role in generating surplus capital to speed up the socioeconomic development of the country (Adem et al. 2018; Hagos et al. 2019).

Achieving food security and end hunger in the face of the ongoing impacts of climate change and variability is at the heart of the sustainable development goals. Climate change and variability are currently the biggest challenge affecting countries where rainfed agriculture is a means of livelihood (Asfaw et al. 2017).

The impact of climate change and variability on agricultural production increases economic and social challenges. Climate change affects agriculture in a different way. Climate extremes and changes in rainfall pattern are already influencing agricultural productivity. Increased tension in households, increased lethargy, poor school performance, and a range of other social ills are the major social impacts mostly reported (Akinseye et al. 2013).

Climate change and variability affect four dimensions of food security, namely, food availability, food accessibility, food utilization, and food system stability. The most direct impact on food security is through changes in food production. Short-term variations are likely to be influenced by extreme weather events that disrupt production cycles (Alemu and Mengistu 2019). Climate change and variability affect food availability through its impacts on economic growth, income distribution, and agricultural demand markets, food prices, and supply chain (Schmidhuber and Tubiello 2007).

Historically, Ethiopia is well known for being prone to extreme events. The rainfall over Ethiopia exhibited high variability (Suryabhagavan 2017), and the country's economy has been affected by long-term changes in both rainfall amount and distribution where the country has witnessed frequent incidents of both excessive and deficient rainfalls (Zeleke et al. 2017). The most common drought periods known are 1957–1958, 1972–1974, 1984–1985, 2002–2003, and 2015–2016 (Markos 1997; Suryabhagavan 2017).

The central highland of Ethiopia, where the study area is located, is a drought- and famine-prone area. The people mainly derive their livelihood from subsistence agriculture, which is characterized by mixed farming system on fragmented land, over utilized land, and affected by erratic rainfall (Alemayehu and Bewket 2016a). Minor droughts are common every year due to climate variability. This brought an impact on the poor performance of the agricultural sector which affects rural livelihoods and food security (Asfaw et al. 2017). The central highland of Ethiopia is already food insecure, and large parts of the Basona Werana district are beneficiaries of the Productive Safety Nets Program (PSNP), which is the major food security program of the government (Alemayehu and Bewket 2017b).

Adaptation enables farmers to reduce vulnerability to adverse effect of climate change and variability. It helps rural communities cope with and adjust adverse consequences to climate change and variability (Deressa et al. 2010; IPCC 2014).

Degefa (2002), Frehiwot (2007), and Messay (2010) investigated the socioeconomic impacts of food security in different regions of Ethiopia. However, the impact of climate change and variability on food security is not addressed in their respective study areas.

Kahsay and Gebremicale (2018) investigated the impact of climate change on household food availability in Tigray region. The findings show that high prevalence of food insecurity has greater consequence among the female households. There are studies addressing the vulnerability context and impacts of climate change covering the central highlands of Ethiopia, which is also part of the present study (Bewket 2009; Alemayehu and Bewket 2016a, b, 2017a, b). This chapter links climate change and variability and food security in Basona Werana district by taking three agro ecological zones. This household level assessment is useful to identify and prioritize food insecure areas and contribute factors for adaptation planning.

Three kebeles (the lowest administrative level in the country) incorporating Kolla (lowland), Weyna Dega (midland), and Dega (highland) agro ecological zones were selected purposively to capture variations in agro ecology (Fig. 1). A total of 123 farmers were selected randomly (53 in Goshebado, 40 in Gudobert, and 30 in Kore Margfiya, respectively). Focus group discussions and key

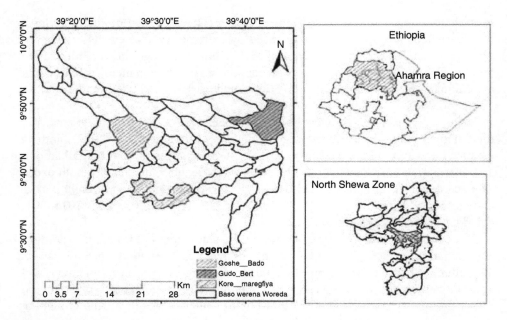

Fig. 1 Location of Basona Werana district, central highlands of Ethiopia

informant interviews were conductive to supplement the quantitative results from the household survey.

The monthly rainfall data used for this study are for 56 points (each representing areas of 10 × 10 km) for the period between 1983 and 2016. The monthly maximum and minimum temperatures are for the same points and girds, but cover the period from 1981 to 2016. All were collected from the National Meteorological Agency of Ethiopia. Rainfall and temperature data were analyzed using coefficient of variation (CV), standardized rainfall anomalies (SRA), and linear regression (LR).

CV is expressed as a percentage and computed as

$$CV = (\sigma/\mu) \tag{1}$$

where CV is the coefficient of variation

σ = is the standard deviation

μ = is the mean rainfall

CV is used to classify the degree of variability of rainfall events. The different CV classes are less (CV <20%), moderate (20%< CV <30%), and high (CV >30%).

Standardized rainfall anomalies are calculated to examine the nature of the trends and determines of the dry and wet years in the period of observation as well as frequency and severity of droughts (Agnew and Chappel 1999; Bewket and Conway 2007).

As shown in Agnew and Chappel (1999), SRA is given as

$$SRA = (Pt - Pm)/\sigma \tag{2}$$

where SRA = standardized rainfall anomaly

Pt = annual rainfall in year t

Pm = is long-term mean annual rainfall over a period of observation and

σ = standard deviation of annual Rainfall over the period of observation

The different drought severity classes are calculated as shown in Agnew and Chappel (1999): extreme drought (SRA < −1.65), severe drought (−1.28 > SRA > −1.65), moderate drought (−0.84 > SRA > −1.28), and no drought (SRA > −0.84).

Linear regression was used to detect changes or trends in rainfall and temperatures. It is given as

$$Y = mx + b \qquad (3)$$

where y is dependent variable, m is the slope, x is independent variable, and b is the intercept.

Surface data were generated from 56 points of gridded monthly rainfall and temperature data (10 km × 10 km) using simple kriging interpolation technique with ArcGIS 10.5.

Household Food Balance Model (HFBM) (sheet) was used to analyses the status of household food security of the sample households in order to classify the households as food secure and insecure (Mesay 2010; Derara and Degfa 2016)

$$NGA = (GPi + GBi + GRi + GPSi) \\ - (HLi + GUi + GSi + GVi + NSi) \qquad (4)$$

where

NGA = Net grain available (quintal/household/year); a quintal = 100Kgs

GPi = Total grain production (quintal/household/year)

GBi = Total grain bought (quintal/household/year)

GSi = Total grain obtained through gift or remittance (quintal/household/year)

GPSi = Total grain obtained from previous year (stock/year/household)

HLi = Post-harvest losses due to grain pests, disasters, thievery (quintal/household/year)

GUi = Quantity of grain reserved for seed (quintal/household/year)

MOi = Amount of marketed output (quintal/household/year)

GVi = Grain given to others within a year (quintal/household/year)

NSi = Grain planned to be left by a household for next season (quintal/household/year)

Descriptive statistical methods such as frequencies, means, and percentages were used to summarize the information on food security issues and climate change perceptions, impacts, and adaptation strategies. Chi-square test is used to test the statistical significance of variations across AEZs.

Impact of Climate Change and Variability on Food Security of Rural Household

Variability and Trends in Rainfall and Temperature

As shown in Table 1, the average annual rainfall of the area for the study period is 1013 mm, with standard deviation of 125 mm and CV of 12%. Annual rainfall varies from 784 mm (2011) to 1227.2 mm (1985). The results of trend analysis of annual and seasonal rainfall are presented in Table 1. Accordingly, annual and seasonal rainfalls show statistically significant decreasing trends except Bega (October–February) rainfall where there is no clear trend. Annual rainfall shows statistically significant decreasing trend 63.58 mm/decade at p = 0.05 level. Belg rainfall also showed statistically significant decreasing trend at p = 0.05 level. Result supports previous studies by Williams et al. (2012) and Kebede (2013) who reported declining trends in Belg rainfall. Kiremt rainfall has shown statistically significant decline of 23.2 mm/decade. Elsewhere in Ethiopia, Kiremt rains support the main cropping season locally known as Meher at p = 0.1 level. The declining trends of Kiremt rainfall will aggravate the food security challenges in the area.

The coefficient of variation revealed that rainfall in the area has low inter-annual variability (12%) (Table 1). June to Kiremt rainfall has large contribution to annual rainfall (72%), with CV of 13%. Belg and Bega seasons have high rainfall variability each having CV of 42.6% and 56.8%, respectively. Belg rainfall contributed 17% of annual rainfall.

Rainfall distribution was highly concentrated in the few months of the year (July to September). July and August are the wettest months. July has the highest monthly rainfall (295 mm) which contributes 30% of annual rainfall. August has 287.5 mm rainfall and contributes about 29%. November and December are the driest months, having 11 mm of rainfall each (Fig. 2).

Analysis of the standard rainfall anomaly is used to show the intensity and frequency of drought and inter-annual variation at various spatiotemporal scales. The rainfall pattern showed a characteristic that a dry year is followed by another 2 or 3 dry years and vis-à-vis for the wet years.

Table 1 Trends of annual and seasonal rainfall

	Mean	LR	Wettest year	Amount rainfall (mm)	Driest year	Amount rain fall (mm)	SD	CV
Annual	1013	−63.58[a]	1985	1227.2	2011	783.8	124.4	12.3
Kirmet	733.8	−23.2[b]	1997	918.9	1987	485.3	98.6	13.4
Belg	182.4	−24.8[a]	1987	437.1	1998	58	77.7	42.6
Bega	87.5	−11.6	1997	237.2	2011	13.3	49.7	56.8

LR = linear trends (mm 10/year)
[a]Significant at 0.05 level
[b]Significant at 0.1 level

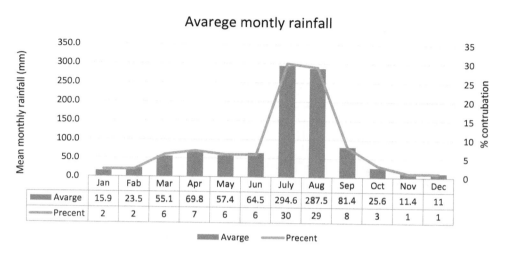

Fig. 2 Monthly rainfall (in mm) and contribution to the annual total (in %)

After the 1984 drought, annual rainfall showed some recovery, but with considerable internal variations. Extreme drought (<-1.65) is observed during the years 1984, 2008, 2009, 2011, and 2015. Severe drought (-1.28 to -1.65) is observed for the years 1989, 1995, and 2005 (Fig. 3). The years 2006 and 2007 show positive anomalies. The year 2006 was a major flood year in the country, which caused loss of life and property in different parts of the country particularly in Dire Dawa and South Omo (Alemayehu and Bewket 2017a). The proportions of negative and positive anomalies account for 61% and 39% of the total observations, respectively. The 1980s and 1990s are wet compared with the 2000s. Close to 23% of observations fall under different drought categories.

At show in Fig. 3, negative and positive Kiremt rainfall anomalies account for 52% and 48% of observations for the period 1983–2016, respectively. Extreme drought was observed in the years 1984 and 1987 and severity drought in 2008 and 2011. Belg rainfall positive anomalies account for 42% of total observation. The extreme drought is observed in 1992, 1997, 1998, 2008, 2009, 2013, and 2014. Belg rainfall anomalies are relatively more positive in the driest decade of the 1980s than the others. Generally, the findings repeat the earlier works of Bewket and Conway (2007), Rosell (2011), Ayalew et al. (2012), and Alemayehu and Bewket (2017a) who reported a similar finding for the central highlands of Ethiopia.

Figure 4 shows the spatial distribution of annual and seasonal rainfall distribution and trends over the period of analysis. Large proportion of the district (42%) receives annual rainfall of between 829 and 886 mm. About 35% of the area in the southern part of the district receives rainfall of between 77 and 829 mm. Close to 33% of the area in the in the northwestern part receives rainfall of between 886 and 955 mm. Most of the grid points of annual rainfall trend fall within -0.84 and -0.04 ranges. Result coincides with a previous study by Alemayehu and Bewket (2017a) on local scale variability, and trends of rainfall and temperature in the central highlands of Ethiopia that reported most of the grid points fall within -0.1 and -0.7 range which

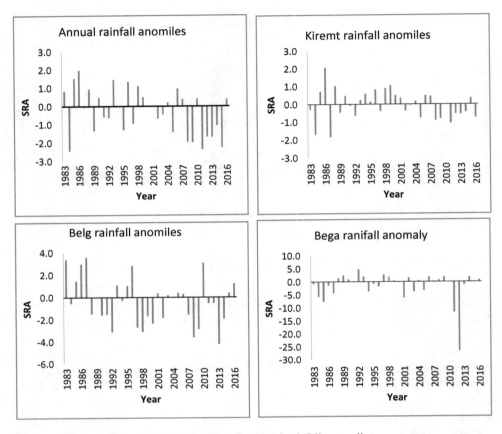

Fig. 3 Temporal variations in the annual and seasonal rainfall anomalies

exhibit strong negative trend of annual rainfall. The implication is that drying trends in the district have adverse consequences on the already poor performance of the agricultural sector in the area. Drying trend of Belg rainfall (−0.4 to −0.02) in the area is observed. The implication is on the Belg season production, which is important for household level food security in the locality (Alemayehu and Bewket 2017a). In the case of Kiremt rainfall (−3 to 2) trends in the upper part of the district is observed. In the middle part of the district (0 to 2), positive trends were observed, whereas the lower part of the district negative trends (−3 to −1).

The mean annual temperature is 13.2 °C. The lowest and highest mean annual temperatures are experienced in 1999 and 2009, respectively. The lowest temperature is 12.1 °C and highest temperature 14.5 °C with a standard deviation of 0.46 and coefficient variation of 0.03. The lowest temperature is −2.1 °C (December) with a standard deviation of 0.8 and coefficient variation of 0.12. The highest temperature is 24.4 °C (February) with a standard deviation of 0.42 and coefficient variation of 0.02. The mean annual minimum temperature ranges from 4.83 °C (2000) to −2.1 ° C (1998), and the long-term mean is 6.3 °C. The mean annual maximum temperature ranges from 23.4 °C (2009) to 15.8 °C (1994), and the long-term mean is 19.7 °C.

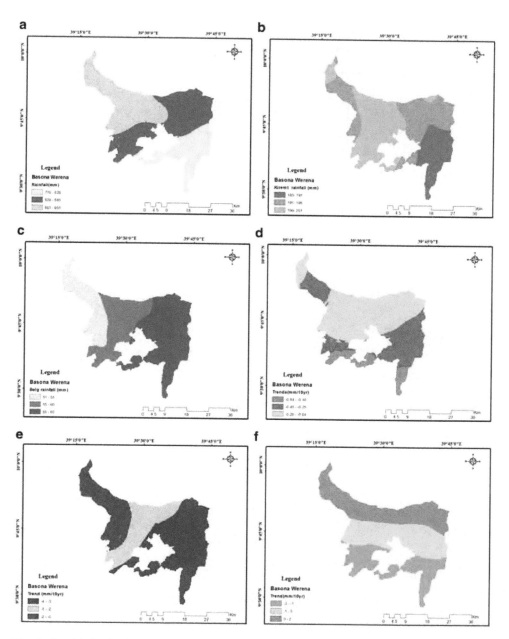

Fig. 4 Spatial distribution of annual rainfall and seasonal rainfall (A, B, C) and annual rainfall and seasonal trends (D, E, F)

The mean annual maximum temperature shows warming trends for the period 1983–2016. The warming trend in the maximum temperature (0.2 °C/decade) is statistically significant at p = 0.05 level.

As show in Table 2, Bega and Belg season's maximum temperature experienced statistically significant increasing trends at p = 0.05 and 0.1 levels, respectively.

Table 2 Annual and seasonal trends in the mean maximum and minimum temperature

	Mean maximum and minimum temperature °C							
	Annual	LT	Kiremt	LT	Bleg	LT	Bega	LT
Maximum temperature °C	19.7	0.02[a]	19.36	−0.1	20.96	0.03[b]	19.6	0.28[a]
Minimum Temperature °C	6.29	0.42[a]	8	−0.015	7.27	0.0114[b]	3.61	0.013

LR = linear trends (0°/10 year)
[a]Significant at 0.05 level
[b]Significant at 0.1 level

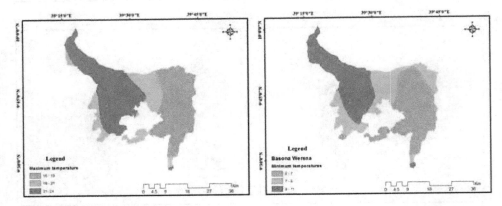

Fig. 5 Spatial distribution of mean maximum (left) and minimum temperatures (right)

Kiremt season's maximum temperature experienced statistically nonsignificant decreasing trend.

The annual minimum temperature shows positive trend. Minimum temperature has experienced statistically significant increasing trend at p = 0.01 level. Kiremt minimum temperature shows declining trend at p = 0.05 level. The trends for the Bega and Belg minimum temperatures are similar with the maximum temperature; statistically significant increasing trends at p = 0.05 and 0.1 levels, respectively, are observed (Table 2). Earlier studies by Conway et al. (2004), NMA (2007), Jury and Funk (2012), Tesso et al. (2012), and Taye and Zewdu (2012) also reported a warming trend of the minimum temperature in their respective study areas and periods.

Figure 5 shows the spatial distribution of mean maximum and minimum temperatures. Both mean maximum and minimum temperatures are higher in the northwestern part of the district and decrease in the southeast. The maximum temperature for northern parts of district ranges from 22 °C to 24 °C and the lowest southeast 10–20 °C, whereas the minimum temperature of the district is lowest in the southeast ranging from 6 °C to 7 °C in the study period.

Rural Household Perceptions of Climate Change and Variability and Their Impacts

Table 3 shows that about 88% of farmers perceived that the climate has been changed. Close to 94% of households in the Kolla agro ecological zone perceived that they have been experiencing climate change and variability. This is followed by Dega agro ecological zone where 93% of households perceived climate change and variability.

> A key informant from Dega agro ecological zone explains: in the last three decades, Belg rains are highly variable and shows declining trends in the surrounding. Belg rains now are non-existent become history to the area. The worst case is that Kiremt rainfall is insufficient due to increasing water demand. Kiremt rainfall starts late and ends early. Coupled with other environmental changes like land degradation and soil erosion, current climate variability challenges food security; not capable of producing enough food for family members.

Results from FGDs also confirmed that rainfall is declining from time to time while temperature is getting warmer. About 67% of households in all agro ecological zones perceived that temperature shows warming trend, while 20% of households perceived that temperature shows declining trends in their locality. Close to 13% of households report no observable change of temperature. Statistically significant mean difference was observed in terms of perception of temperature across agro ecological zones at p = 0.1 level.

Regarding changes in rainfall patterns, close to 59% of households reported that rainfall shows decreasing trends. The other 35% of households observed changes in the timing of rainfall. Farmers in the Kolla agro ecological zone reported declining trends in rainfall (73% of households). About 53% and 52% of households from

Table 3 Rural household perceptions of climate change and variability and their impacts

Climate change/variability	Agro ecological zone			Mean	X^2
	Dega	Kolla	Weyna Dega		
Yes	93	94	70	88	11.819[a]
No	8	6	30	12	
Temperature pattern					
Increased	65	71	64	67	8.323[b]
Decreased	27	20	11	20	
No observable change	8	9	25	13	
Rainfall perceived					
Decrease in rainfall amount	52	73	53	59	5.111
Change in timing of rain	43	24	37	35	
No change in rainfall	5	3	11	6	

[a]Significant at 0.05 level
[b]Significant at 0.1 level

Weyena Dega and Dega agro ecological zone perceived declining of rainfall trends, respectively.

Rural households' perception of changes in rainfall and temperature corroborates the results of metrological data analysis. They perceived reduced rainfall and warming temperatures. Analysis of climate data revealed declining trends of annual and seasonal rainfalls. Similarly, warming of the minimum and maximum temperatures is observed from analysis of climate data.

Regarding the cause of climate change, about 39% of households reported human interventions caused climate change. Close to 11% of households in the three agro ecological zones reported natural factors caused climate change. About 40% and 36% of households in the Dega and Kolla agro ecological zones attributed climate change with religious factors. Statistically significant mean difference was observed in terms of perceived causes of climate change across agro ecological zones at $p = 0.01$ level. Results of FGD participants and KIIs reported the effect of deforestation for different purposes is the main cause of climate change.

Impacts of Climate Change and Variability on Food Security of Rural Households

As show in Table 4, majority of the households (88%) reported changes in rainfall pattern as the main challenge of their livelihoods. The onset and offset periods of rainfall are unpredictable and affect sowing and harvesting periods. Prevalence of pest and diseases was mentioned by 82% of the households. The effect of flood is reported by 48% of the households. Snow as a challenge to food security is reported by 65% of the households. For example, months from October to December are the coldest months in the area. Thus, matured crops are more likely to be affected by frost. Drought has several negative consequences in the area; it causes extensive damages to crops and loss of agricultural production. Close to 67% of the households replied that drought was the major climate-related factor challenging agricultural productivity in the area (Table 4).

Table 4 Impact of climate change and variability on livelihood

Major climate elements	Agro ecological zone			Mean	X^2
	Dega	Kolla	Weyna Dega		
Temperature increased	79	96	87	87	6.906[a]
Change in rainfall	79	97	87	88	4.758
Drought	58	73	67	66	9.883[b]
Flood	58	34	50	48	2.189
Snow	60	71	63	65	1.361
Pests and diseases	88	76	83	82	10.182[b]

[a]Significant at 0.05 level
[b]Significant 0.01 level

Table 5 Correlation between climate change and variability and food security

		Temperature change	Rainfall variability	Flood	Drought	Frost
Food security	Pearson Correlation	−0.182	−0.157	0.231	−0.150	−0.174
	Sig.(2-tailed)	0.044	0.083	0.010	0.098	0.055
	N	123	123	123	123	123

Statistically significant mean difference was observed in terms of perceived impacts of climate change and variability on livelihoods across agro ecological zones.

Bivariate correlation was used to analyze the association between food security and climate change and variability over the last two to three decades. The food security status of a household is determined by the rate of temperature and rainfall changes, occurrence of flood, drought, and frost which have adverse negative impacts on crop and animal productivity thereby affecting the food security status of rural households (Table 5).

Results of the bivariate correlation showed that food security has negative correlation with temperature change at p = 0.05 level. Similarly, rainfall variability and occurrence of drought and frost have negative correlations at p = 0.1 levels. Food security has positive correlation with occurrence of flood at p = 0.05 level.

Household Food Security Status

HFBM was used to determine household food security status. The result of the HFBM reveals that 62% of households are food insecure and failed to satisfy their daily minimum requirement recommended for their households (2100 Kcal/adul. equ), while 38% of the households are food secured. At agro ecological zone level, food insecurity is highest in the Dega (67% of the households) followed by Kolla (60% of the households), whereas majority of food secured households are found in Weyna Dega agro ecological zone (43% of the households). This is because of the fact that topographic and climate factors are conducive for crop production in the Weyna Dega agro ecological zone compared to others (Alemayehu and Bewket 2016b).

There are different constraints that hinder agricultural productivity and then induce food insecurity. As shown in Table 6, almost all households (99%) replied that poor soil fertility was the main factor reducing productivity thereby causing food insecurity. This is followed by climate variability (97% of the households). About 95% and 93% of the households reflected that farmland shortage and low non-farm income are among the major causes for food insecurity, respectively. Farmers felt that insect, pests, and weeds caused food insecurity. Statistically significant variation in terms of agro ecological zone is observed among the causes of food insecurity at P = 0.01 level.

Table 6 Cause of food insecurity

Cause of food insecurity	Agro ecological zone			Mean	X^2
	Dega	Kolla	Weyna Dega		
Traditional farm practices	98	98	96	97	1.230
Land fragmentation	100	83	63	82	16.962[a]
Shortage of labor	95	76	90	87	7.641[b]
Poor soil fertility	100	96	100	99	2.685
Climate variability(rainfall and temperature	100	90	100	97	6.884[b]
Prevalence of animal disease	70	88	67	75	5.175[c]
Deforestation	97	77	73	82	9.111[b]
Farm land shortage	100	98	88	95	8.948[b]
Drought	85	57	60	67	9.065[b]
Poor access credit	88	66	37	64	6.701[c]
Low non in farm income	89	97	93	93	3.613
Shortage of grazing land	88	72	90	83	5.719[b]
Low access market	57	80	40	59	11.936[a]

[a]Significant 0.01 level
[b]Significant at 0.05 level
[c]Significant at 0.1 level

Shortage of labor, climate variability, deforestation, farm land shortage, and shortage of grazing land as cause of food insecurity show statistical difference across agro ecological zone at p = 0.05 level. While lack of credit access and prevalence of animal diseases show significant difference at p = 0.1 level. The office of Agriculture and Natural Resources of the Amhara Regional State reported that about 15–26% of post-harvest production loss was observed. Coupled with the already low production, the post-harvest loss further affected household food security through diminishing the amount of available food reserve.

This result supported a previous study by Tilaye (2004) in Amhara Region of Ethiopia which identified different factors that cause food insecurity by deteriorating the food production capacity of households. These included drought, soil erosion, land fragmentation, population and pressure poor farming technology, and high labor west age as major causes of food insecurity.

In response, households used different coping and adaptation strategies to mitigate the adverse effects of climate change and variability on food security. About nine types of adaptation measures are identified. These are soil and water conservation, changing planting date, use of fertilizer, planting tree, animal fattening, livelihood diversification, irrigation, improved seed, and crop diversification. The majority of farmers (99%) used soil and water conservation practices as adaptation to the impact climate change variability on food security. This is followed by changing crop planting dates (98% of households). While the least strategies are irrigation (21%) and animal fattening (59%). Similarly, changing consumption pattern and sell of livestock product (butter, milk, and cheese) are the dominant coping strategies used by 88% and 83% of the households, respectively. While migration and purchase of food by credit are the least coping strategies used by 21% and 37% of the households, respectively.

References

Adem M, Tadele E, Mossie H, Ayenalem M (2018) Income diversification and food security situation in Ethiopia: a review study. Food Sci Technol 4:1–17

Agnew C, Chappel A (1999) Drought in the Sahel. Geo J 48:299–311

Akinseye F, Ajayi V, Oladitan T (2013) Assessing the impacts of climate variability on crop yield over Sudano-Sahelian zone in Nigeria. Int J Agric Sci 1(7):91–98

Alemayehu A, Bewket W (2016a) Local climate variability and crop production in the central highlands of Ethiopia. Environ Dev 19:36–48

Alemayehu A, Bewket W (2016b) Vulnerability of smallholder farmers' to climate change and variability in the central highlands of Ethiopia. Ethiopian Journal of the Social Sciences and Humanities 12(2):1–24

Alemayehu A, Bewket W (2017a) Smallholder farmers' coping and adaptation strategies to climate change and variability in the central highlands of Ethiopia. Local Environ 22(7):825–839

Alemayehu A, Bewket W (2017b) Determinants of smallholder farmers' choice of coping and adaptation strategies to climate change and variability in the central highlands of Ethiopia. Environ Dev 24:77–85

Alemayehu A, Bewket W (2017c) Local spatiotemporal variability and trends in rainfall and temperature in the central highlands of Ethiopia. Geogr Ann Ser B 99(2):85–101

Alemu T, Mengistu A (2019) Impacts of climate change on food security and its adaptation and mitigation options in Ethiopia: a review. In: International conference on impact of El Niño on biodiversity, agriculture, and food security, 23–24 February 2017 Haramaya University, Ethiopia, p 75

Asfaw A, Simane B, Hassen A, Bantider A (2017) Determinants of non-farm livelihood diversification: evidence from rain fed-dependent smallholder farmers in northcentral Ethiopia (Woleka sub-basin). Dev Stud Res 4(1):22–36

Ayalew D, Tesfaye K, Mamo G, Yitaferu B, Bayu W (2012) Variability of rainfall and its current trend in Amhara region, Ethiopia. Afr J Agric Res 7(10):1475–1486

Belay A, Recha J, Woldeamanuel T, Morton (2017) Smallholder farmers' adaptation to climate change and determinants of their adaptation decisions in the Central Rift Valley of Ethiopia. Agric Food Secur 6(1):24

Bewket W (2009) Rainfall variability and crop production in Ethiopia: case study in the Amhara region. In: Ege S, Aspen H, Teferra B, Bekele S (eds) Proceedings of the 16th international conference of Ethiopian studies. Norwegian University of Science and Technology, Trondheim, Norway

Bewket W, Conway D (2007) A note on the temporal and spatial variability of rainfall in the drought-prone Amhara region of Ethiopia. Int J Climatol 27:1467–1477

Conway D, Mould C, Bewket W (2004) Over one century of rainfall and temperature observations in Addis Ababa. Ethiop Int J Climatol 24:77–91

Degefa T (2002) Household seasonal food insecurity in Oromiya Zone, Ethiopia: S.S research report series no 26, OSSREA, A.A

Derara F, Degfa T (2016) Household food security situation in Central Oromia, Ethiopia: a case study from Becho Wereda in Southwest Shewa Zone. Glob J Hum Soc Sci Res 16 (2):1–16

Deressa T, Hassan R, Tekie A, Mahmud Y, Ringler C (2010) Analyzing the determinants of farmers choice of adaptation methods and perceptions of climate change in the Nile Basin of Ethiopia. Sustainable solutions for ending hunger and poverty. IFPRI discussion paper 00798. IFPRI, Washington DC

Frehiwot F (2007) Food insecurity and its determinants in rural households in Amhara Region. Msc thesis, Department of Economics, Faculty of Business and Economics, School of Graduate Studies, Addis Ababa University, Ethiopia

Hagos A, Dibaba R, Bekele A, Alemu D (2019) Determinants of Market Participation among Smallholder Mango Producers in Assosa Zone of Benishangul Gumuz Region in Ethiopia. Int J Fruit Sci 20(3): 323–349

Intergovernmental Panel on Climate Change (IPCC) (2014) Climate change: the scientific basis: contribution of working group I to the third assessment report. Cambridge University Press, Cambridge

Jury R, Funk C (2012) Climatic trends over Ethiopia: regional signals and drivers. Int J Climatol. http://onlinelibrary.wiley.com/doi/10.1002/joc.3560

Kahsay S, Gebremicale D (2018) Impact of climate variability on food availability in Tigray, Ethiopia. J Agric Food Secur 7(6):1–9

Kebede G (2013) Spatial and temporal uncertainty of rainfall in arid and semi-arid areas of Ethiopia. Sci Technol Arts Res J 2(4):106–113

Markos E (1997) Demographic responses to ecological degradation and food insecurity: drought prone areas in northern Ethiopia. Amsterdam. Thesis Publishers Ph.D. dissertation

Mesay M (2010) Food security attainment role of urban agriculture: a case study from Adama Town, Central Ethiopia. J Sustain Dev Afr 12(3):223–250

Mesike S, Esekhade T (2014) Rainfall variability and rubber production in Nigeria. Afr J Environ Sci Technol 8(1):54–57

Mongi H, Majule E, Lyimo G (2010) Vulnerability and adaptation of rain fed agriculture to climate change and variability in semi-arid Tanzania. Afr J Environ Sci Technol 4(6)

National Meteorological Agency (NMA) (2007) Climate change: National Adaptation Program of Action (NAPA) of Ethiopia. Ministry of Water Resources, Addis Ababa

Rosell S (2011) Regional perspective on rainfall change and variability in the central highlands of Ethiopia, 1978–2007. Appl Geogr 31:329–338

Schmidhuber J, Tubiello F (2007) Global food security under climate change. Proc Natl Acad Sci 104(50):19703–19708

Suryabhagavan K (2017) GIS-based climate variability and drought characterization in Ethiopia over three decades. Weather Clim Extrem 15:11–23

Taye M, Zewdu F (2012) Spatio-temporal variability and trend of rainfall and temperature in Western Amhara, Ethiopia: a GIS approach. Glo Adv Res J Geogr Reg Plann 1(4):65–82

Tesso G, Emana B, Ketema M (2012) A time series analysis of climate variability and its impacts on food production in North Shewa zone in Ethiopia. Afr Crop Sci J 20(2):261–274

Tilaye T (2004) Food insecurity: extent, determinants and household copping mechanisms in Gera Keya Woreda, Amhara region. MA thesis, Addis Ababa University

Williams P, Funk C, Michaelsen J, Rauscher S, Robertson I, Wils T, Koprowski M, Eshetu Z, Loader N (2012) Recent summer precipitation trends in the greater horn of Africa and the emerging role of Indian Ocean Sea surface temperature. Clim Dyn 39:2307–2328. https://doi.org/10.1007/s00382-011-1222-y

Zeleke T, Giorgi F, Diro GT, Zaitchik B (2017) Trend and periodicity of drought over Ethiopia. Int J Climatol 37:4733–4748

5

Climate-Induced Food Crisis in Africa: Integrating Policy and Adaptation

David O. Chiawo and Verrah A. Otiende

Contents

Abstract

Climate change threatens development and economic growth in Africa. It increases risks for individuals and governments with unprecedented negative impacts on agriculture. Specifically, climate change presents a major threat to food security in Africa for the long term due to the low adaptive capacity to deal with successive climate shocks. There is a need for greater awareness of the trends of food crisis patterns and adaptive initiatives. The objective of this chapter was to analyze the trends of the food crisis in Africa within the past 10 years and adaptive initiatives. Quantitative data analyzed for food security indicators were

D. O. Chiawo (✉)
Strathmore University, Nairobi, Kenya
e-mail: dchiawo@strathmore.edu

V. A. Otiende
Pan African University Institute for Basic Sciences Technology and Innovation, Nairobi, Kenya

obtained from the United Nations Food and Agriculture Organization (FAO) and World Development Indicators (WDI) available at the Environment and Climate Change data portal. Policy and adaptation measures related to climate change were reviewed in 26 countries in Africa, with the view to highlight their integrative nature in enhancing food security. High prevalence of undernourishment was observed in six countries, all in sub-Saharan Africa (SSA) including Chad, Liberia, Central African Republic, The Democratic Republic of the Congo, Zambia, and Zimbabwe. Countries with a high land acreage under cereal production recorded reduced undernourishment. Niger demonstrated effective adaptation for food security by registering the highest crop production index in extreme climate variability. However, Kenya appears to be the most predisposed by registering both high climate variability and below average crop production index. It is observed that diversification and technology adoption are key strategies applied across the countries for adaptation. However, the uptake of technology by smallholder farmers is still low across many countries in SSA.

Keywords

Climate change adaptation · Climate change adaptation policy · Food security · Smallholder farmers · Africa

Introduction

Climate change threatens development and economic growth in Africa. It will increase risks for individuals and governments with unprecedented negative impacts across all sectors of the economy. Specifically, climate change presents a major threat to food security in Africa for the coming decade due to low adaptive capacity to deal with successive climate shocks, despite the trend of the rapid pace of population growth. Population growth in sub-Saharan Africa is likely to experience by far the most rapid relative expansion more than doubling to 2.0 billion by 2050 (Bongaarts 2009). The positive population growth is attributed to technology transitions in health care; however, it is expected to generate increased demand for agricultural land and forest products (Brandt et al. 2017), calling for a matching assemblage of technology and policies to enhance food security to limit the possible impact to the environment.

An overwhelming majority of hungry people are found in developing countries, particularly in Africa (Abegunde et al. 2019). With increasing recognition of present and future impacts of climate change, the United Nations Framework Convention on Climate Change (UNFCC 2006) has identified the poorest people living in developing countries as the most vulnerable due to their dependence on natural resources and rain-fed agriculture for survival. However, agriculture remains the dominant employer of rural residents in Africa (e.g., Kalungu and Leal Filho 2018; Belay et al. 2017; Mutunga et al. 2017; Cobbinah and Anane 2016). Smallholder farming creates opportunities for an estimated 175 million people in Africa, of which about 70% are women (AGRA 2014). Across the world, smallholder farmers are considered to be disproportionately vulnerable to climate change because changes in temperature, rainfall, and the frequency or intensity of extreme weather events directly affect their crop and animal productivity (Vignola et al. 2015).

The vulnerability of smallholder farmers is a concern because they represent 85% of the world's farmers and provide more than 80% of the food consumed in the developing world; therefore, what happens to smallholder farmers will have significant social, economic, and environmental consequences globally (Vignola et al. 2015). By their sheer numbers and with optimal productivity, they would have a huge impact in addressing food insecurity (Abegunde et al. 2019), but because of low resource capacity, inadequate skills, and lack of enabling policies, they remain susceptible particularly to climate variability in Africa. Climate change adaptation has become necessary to alleviate the impacts of extreme weather events, especially on rural farmers whose primary livelihood is climate dependent (Cobbinah and Anane 2016).

There is a looming food security crisis in Africa for the next decade due to limited adaptive approaches to extreme climatic conditions of drought, floods, and high temperatures due to climate change (Abegunde et al. 2019). IPCC special report suggested that an increase by 2 °C could exacerbate food crises with crops under rain-fed agriculture dropping by 50% in some African countries by 2020 (IPCC 2018). The current and future impacts of climate change have attracted increasing scholarly work on climate change adaptation (Cobbinah and Anane 2016; Burnham and Ma 2016; Lasco et al. 2014) and its widespread incorporation into national policy and international dialogue (IPCC 2007, 2013). Responding to the effects of climate change requires the continuous development of new adaptive initiatives and improvement of the existing ones and enhancing their widespread adoption by smallholder farmers (Kalungu and Leal Filho 2018). Pursuing appropriate adaptation initiatives in line with realities in Africa is thus vital. A recent study highlighted a positive link between public response and the capacity of farmers to adapt to climate change in sub-Saharan Africa (Abegunde et al. 2019). However, a gap exists in critical review of the role of existing climate change adaptation policies in enhancing the adoption of the initiatives. The objective of this chapter is to analyze the spatiotemporal trends of the food crisis in Africa and adaptive approaches with the view to model a framework for climate change adaptation pathway for Africa.

Definition of Key Concepts

This chapter uses the definition of food security by the United Nations Food and Agriculture Organization (FAO) as food is available at all times; that all persons have means of access to it; that is nutritionally adequate in terms of quantity, quality, and variety; and that it is acceptable within the given culture (Silvestri et al. 2015; Conceição et al. 2016; Luximon and Nowbuth 2010). The chapter defines climate change as a change of climate that is attributed directly or indirectly to human activity that alters the composition of the global atmosphere that is in addition to natural climate variability observed over comparable periods (UNFCC 2006). The chapter adopts the meaning of climate change adaptation (Mutunga et al. 2017) in that adaptation to climate change is the adjustment in natural or human systems in response to actual or expected climatic stimuli or their effects, which moderates harm or exploit beneficial opportunities.

Methods and Research Design

The analysis was based on four World Bank food security indicators including average prevalence of undernourishment, land under cereal production, extreme climate change variability, and crop production index. Dataset was obtained from World Development Indicators (WDI) (WDI available at Environment and Climate Change data portal http://africaclimate.opendataforafrica.org/kaoatif/world-development-indicators-wdi). R was for spatial analysis (R Core Team 2020) using shapefile for 52 African countries from Map Library (Map Library is available at http://www.maplibrary.org/library/stacks/Africa/index.htm). Both dataset and shapefile used are licensed under CC BY 4 and are open for public access (Publicly available for any use according to open data standards and licenses under the Creative Commons Attribution 4.0 International license (CC-BY 4.0)). Adaptive and policy strategies applied across Africa were reviewed at local, national, and regional scales. Countries were selected based on the undernourished index and vulnerability to climate change. Spatial patterns were analyzed for undernourishment, land under cereal production, crop production index, and climate change variability in all the 52 African countries according to the shapefile. The spatial patterns were used to identify the countries for review of adaptation and policy strategies. High undernourishment prevalence countries above 40% and low prevalence countries below 10% were selected for review of adaptation and policy strategies for comparison and criticism. Additional counties for review of adaptation and policy were selected based on vulnerability to climate change, low or high crop production index, and landmass under cereal production (Figs. 1 and 2). A total of 26 countries were reviewed across Africa with a larger proportion in SSA. The approach presents a conceptual model integrating climate change adaptation and policy initiatives to illustrate the effective pathway to climate change adaptation in Africa for food security.

Patterns of Climate-Induced Food Crisis and Adaptation in Africa

This section is structured in three parts: the first part presents the results of spatial analysis of food security in Africa for the last 10 years, adaptation initiatives in the second part, and policy initiatives in the third part.

Spatial Trends of Food Crisis in Africa

The prevalence of undernourishment (PoU) is defined by FAO as an estimate of the proportion of the population whose habitual food consumption is insufficient to

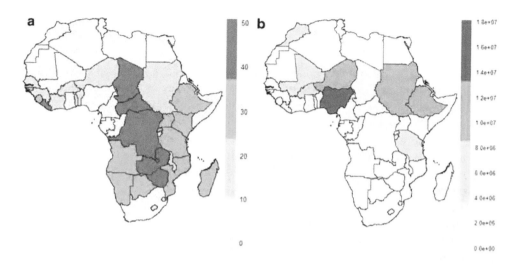

Fig. 1 Spatial patterns of climate-induced-food crisis in Africa. (**a**) Average prevalence of undernourishment (% of the population) in Africa for the period 2009–2017. (**b**) Land under cereal production (hectares) for the period 2009–2017 (Authors own)

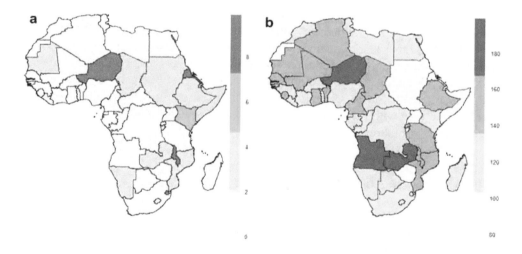

Fig. 2 Spatial patterns of extreme climate change variability and crop production index in Africa. (**a**): Droughts, floods, extreme temperatures (% of the population). (**b**): Crop production index for the period 2009–2016 (Authors own)

provide the dietary energy levels that are required to maintain a normal active and healthy life. The spatial trends of undernourishment in the last 10 years reveal a high prevalence of 40–50% in six countries (6), all in sub-Sahara Africa (SSA) including Chad in the Sahel and Central Africa, Liberia in West Africa, Central African Republic and the Democratic Republic of the Congo in Central Africa, and Zambia and Zimbabwe in Sothern Africa. Low prevalence is observed in North Africa with all countries registering below 10% of undernourishment (Fig. 1a).

Countries with a high land acreage under cereal production recorded reduced undernourishment (Fig. 1a). Nigeria had the largest landmass under cereal production in the last 10 years with a peak of 18 million hectares (Fig. 1b). Other countries above the average of landmass under cereal include Sudan, Ethiopia, Tanzania, and Niger. It is predicted that a possible cross-border trade policy or intercountry relationship between neighboring countries could be influencing the production of cereals, e.g., Sudan and Ethiopia and Niger, Nigeria, Burkina Faso, and Mali. Most countries in Southern Africa have lower than average landmass under cereal production. Although South Africa has below average landmass under cereals, the undernourishment prevalence was still low for the period of analysis, a pointer toward effective diversification or technology adoption in climate change adaptation (Fig.1b).

High incidences of extreme conditions due to climate change including droughts, floods, and extreme temperatures are high in four countries that experienced the highest extreme climate change variability in the last 10 years including Niger, Eritrea, Malawi, and Swaziland. However, these countries registered average to high crop production index an indication of effective climate change adaptation. Kenya appears to be more predisposed to extreme climate change variability in East Africa (Fig. 2a). The leading countries in Africa in crop production are Niger, Angola, and Zambia. A good example of how effective climate change adaptation can lead to food security is Niger, which experiences the highest in extreme climate variability, but registers the highest crop production index. It is important to note that Nigeria had a higher landmass under cereal production (Figs. 1b and 2b), but Niger registered a higher crop production index in the same period pointing toward possibilities of application of multiple adaptive initiatives beyond increasing acreage. Sudan, South Sudan, Eritrea, Somalia, Zimbabwe, and Western Sahara are African countries with the lowest crop production index despite experiencing average and below of extreme climate change variability (Fig. 2).

Climate Change Adaptation Initiatives in Africa

Scientific evidence indicates that the earth's climate is rapidly changing, owing to increases in greenhouse gas emissions leading to raised average temperature and altered the amount and distribution of rainfall globally (Dasgupta et al. 2014). There is growing evidence that the effects of climate change are expected to be greater than the global average in sub-Saharan Africa due to projections of a warming trend characterized by the frequent occurrence of extreme heat events with a 4 °C warming scenario (Serdeczny et al. 2016), increasing aridity, and a decline in rainfall (Belay et al. 2017). Sub-Saharan Africa is particularly vulnerable to these climatic changes because of overdependence on rainfed agriculture on which the livelihoods of a large proportion of the region's population currently depend (Serdeczny et al. 2016).

There are reports of higher drought risks; extreme weather events such as floods, pests, and high temperatures; and diseases faced by farmers in recent years, exacerbating the food crisis in Africa (Mutunga et al. 2017; IPCC 2018). There is evidence from the literature that the current impacts are more severe in rural and smallholder farming communities (Cobbinah and Anane 2016) especially in Africa. This presents the logic that an adaptation initiative is promising only if it fulfills the objective of increasing food security, increases resilience, and is adaptive for smallholder farmers (Descheemaeker et al. 2016). Therefore, adaptation is deemed more important in Africa than mitigation (Descheemaeker et al. 2016). There is a need to strengthen adaptive strategies to ensure food security in Africa. Our review categorizes climate change adaptation in Africa under nine (9) key thematic areas including *diversification, ecosystem-based adaptation, climate advisories and information, collective action, technological interventions and innovation, credit facilities, change of patterns, insurance, and capacity building.*

Diversification

Different types of diversification initiatives are applied in Africa including sector diversification, agriculture product diversification and crop diversification, tree-based diversification, and livelihood diversification to adapt to climate change. Diversification provides smallholder farmers with a diversity of diet and improves their income and nutrition security, particularly in sub-Saharan Africa where rainfed agriculture is rampant. For example, Mauritius diversified agriculture with the rapid expansion of industrialization for exports in textile and the development of tourism to promote sector diversification and crop diversification. Also, the Government of Mauritius widened access to lands to be leased out to small planters specifically for agriculture product diversification (Luximon and Nowbuth 2010). While in Mauritius the initiatives to diversify agricultural products seem to be incentivized by the Government, in Kenya, they are largely driven by smallholder farmers through off-farm employment, leasing of land (Kalungu and Leal Filho 2018), mixed cropping and livestock farming (Mutunga et al. 2017), and introducing new crops that are drought resistant or switch to a different variety of the same crop (Crick et al. 2018).

In Morocco, farmers practice crop diversification with intercropped vegetables and tree-based diversification using high-value crop trees (HVC), e.g., drought-tolerant species of fig, almond, and pomegranate and olive varieties (Kmoch et al. 2018). In Ethiopia, farmers practice crop diversification, agricultural product diversification by increasing the integration of crops with livestock, and tree planting. However, reducing the number of meals is common for livelihood diversification to adapt to limited food supply (Belay et al. 2017). In Tanzania, farmers diversify agricultural products by growing vegetables in the offseason and sell livestock during drought, while in South Africa, livelihood diversification and focus on livestock are common in case of crop failure. In Mozambique, farmers diversify to other food sources to meet the dietary needs and shift focus to other means of livelihood beyond agriculture. Farmers in Ghana diversify to enhance food production through multiple cropping, combine improved and local crop varieties, and shift from palm oil to maize and sweet potatoes (Burnham and Ma 2016). Focus on cultivating cash crop (cashew) and engaging in non-farm-based activities, e.g., business, are also common in Ghana. Cashew cultivation is viewed by small landholdings in Ghana as a better livelihood option compared to non-farm-based activities (Cobbinah and Anane 2016). To adapt livestock husbandry to the effects of the extreme of climate change, rural communities in sub-Saharan Africa have diversified to goats instead of cattle and sturdy African breeds instead of more productive crossbreeds (Descheemaeker et al. 2016). In northwest Tunisia, small ruminant farms are better able to adopt the organic farming system and to adapt to warming or precipitation increases by switching to heat-tolerant animals like goats or crops such as Sulla (Khaldi et al. 2012).

Crop diversification together with the diversification of income from multiple sources (cash and in-kind, farm, livestock, crops, and off-farm) is considered to be key "buffer strategies" smallholder farmers pursue to deal with risks of climate change to agrarian environments in Africa (Silvestri et al. 2015; Cobbinah and Anane 2016; Burnham and Ma 2016; Belay et al. 2017; Khacheba et al. 2018; Mutunga et al. 2017). The effectiveness of diversification initiative has been observed in food-secure female-headed households who allocate twice as much land to HVC, e.g., fruits, such as mango and oranges, and then their food-insecure counterparts (Silvestri et al. 2015). This concept in Malawi has demonstrated a gender perspective in climate change adaptation that could be a wake-up call to many African countries to enhance women's adaptive capacity and resilience to climate change (Nyasimi et al. 2014). Indigenous fruit tree production for improved nutrition and income is successful in Malawi, Zambia, and Zimbabwe. Fruit consumption has clear health and nutrition benefits such as providing essential micronutrients and protecting against chronic diseases. Even though fruit consumption in Africa is low, fruit trees fulfill a vital role in contributing to food security, because the fruit is consumed seasonally at a time when households run out of food (Kiptot et al. 2014). There is a need to extensively explore diversification through fruit trees across many African countries to enhance food security.

Ecosystem-Based Adaptation

Ecosystem-based adaptation is the use of biodiversity and ecosystem services as part of an overall adaptation strategy to help farmers adapt to the adverse effects of climate change (Vignola et al. 2015). There is a rapidly growing interest in ecosystem-based adaptation for its potential social, environmental, and economic benefits (e.g., Cerdán et al. 2012; Namirembe et al. 2014; Vignola et al. 2015). Agroforestry provides environmental and social benefits as part of farming livelihoods. In most documented cases of successful agroforestry establishment, tree-based systems are more productive, more sustainable, and more attuned to people's cultural or material needs than treeless alternatives (Mbow et al. 2014). Trees or shrubs on farms and forest-agriculture integration are gaining preference among farmers in Africa. Despite the productivity of agroforestry systems, it has not been fully adopted in many African countries, e.g., Kenya's intention to achieve a 10% target of tree cover on farmers' acreage is yet to be realized (Oloo 2013). There is a need for more insights into the productivity and environmental performance of agroforestry systems across Africa (Mbow et al. 2014). Agroforestry with leguminous fodder plants to support food requirements for livestock during extreme weather is gaining attention among farmers in sub-Saharan Africa (Cobbinah and Anane 2016). In

Tanzania, farmers focus on planting trees as hedgerows, while in Nigeria on-farm trees are used to shade crops and animals (Burnham and Ma 2016). Tree-based agroforestry is also popular in Morocco where farmers use bi- or triennial trees like olives to increase tree cover on-farm (Kmoch et al. 2018).

Improved tree fallows using leguminous trees are encouraged in maize fields in Malawi, Zambia, and Zimbabwe as low cash-input agroforestry practices to restore soil fertility (Keil et al. 2005). Studies in semi-arid areas in Tanzania showed that agroforestry tree species producing high-quality litter could enhance post-fallow soil nutrient availability and crop yields through mineralization of soil organic matter and green manure (Kimaro et al. 2008). In Mauritius, farmers promote the use of organic farming, using organic manure which is more resistant to drought and also reduces the risk of floods (Luximon and Nowbuth 2010), and in Kenya, farmers using locally available organic matter prepare organic manure (Oloo 2013). Farmers in Morocco adapt to the effects of climate change like floods by encouraging conservation agriculture through minimum tillage and increase of soil cover (Kmoch et al. 2018). Minimum tillage and crop rotation were observed to increase farm maize productivity by about 26–38% for minimum tillage and 21–24% for crop rotation in Zambia (Kuntashula et al. 2014). In sub-Saharan Africa rangeland, farmers practice grazing and rangeland management to enhance manure collection, storage, and application (Descheemaeker et al. 2016).

Planting cover crops and the use of manure for soil conservation to prevent soil erosion and to improve soil fertility respectively are common among small-holder farmers in Tanzania, Kenya, and Ethiopia (Mutunga et al. 2017; Belay et al. 2017; Burnham and Ma 2016). Farmers in Ethiopia are observed to practice both soil and water conservation (Belay et al. 2017). In Burkina Faso, farmers arrange stone lines on-farm and in the field to capture surface runoff for micro-water harvesting and to prevent soil erosion due to flash floods, and in South Africa, farmers build bunds along contour lines to slow down water runoff (Burnham and Ma 2016).

Technology and Innovation Adoption

The adoption of appropriate technologies in small-scale farming is an important response to the effects of climate change and variability, especially in Africa (Kalungu and Leal Filho 2018). Farmers largely apply integrated pest management as a means of pest control (Luximon and Nowbuth 2010). Literature indicates a wide use of technological intervention to enhance adapting dairy farming in Tunisia, e.g., farmers in Tunisia build up fodder reserves in favorable years, increase concentrate distribution, manage nutritional requirements, and promote drinking water ad libitum (Hajer et al. 2018). They are keen on the choice of breed and maximize the use of agri-byproducts as feeds. To improve productivity, farmers in Tunisia enhance irrigation potential at the farm level. Farmers also adjust calving periods by concentrating them within favorable feeding periods (Belay et al. 2017). Building fodder reserves for animals is also documented in Burkina Faso (Burnham and Ma 2016). Farmers in Burkina Faso also overcome the shock of drought by producing dry season vegetables using irrigation and the use of improved seeds.

In Kenya, farmers in semi-arid regions apply a combination of technological interventions under Framing Technologies, e.g., soil fertility management, irrigation soil conservation, and water management row planting and terracing. There is high-level awareness on the use of hybrid crop varieties, pesticides, and changing crop variety depending on climate variability (Mutunga et al. 2017; Kalungu and Leal Filho 2018). Planting short seasoned crops and resistant crop variety and feed preservation for farm animals have also been recognized as key adaptive interventions among Kenya farmers (Oloo 2013). Rotation cropping of cereals and legumes and irrigated fruit trees are common in commercially orientated farming systems in Morocco, with agro-silviculture (growing of trees and agriculture crops together in the same lands at the same time) dominating all irrigated farms (Kmoch et al. 2018). There is input intensification in Ethiopia to enhance farm-level productivity (Belay et al. 2017), and in Tanzania, farmers use well irrigation and plant quick maturing and drought-resistant varieties (Burnham and Ma 2016). Framers in South Africa increase distances between crop rows and practice irrigation. Farmers in South Africa also plant late-maturing fruit trees to enhance food availability during low harvests (Burnham and Ma 2016). In Ghana, farmers plant high yield varieties (Belay et al. 2017). In other countries relay cropping and mixed intercropping are practiced. By providing nutrients to crops, these technologies can potentially help farmers improve their soils and incomes, thereby improving food security in Malawi, Zambia, and Zimbabwe (Kiptot et al. 2014).

Farmers in many African countries aim at enhancing farm-level productivity by planting early maturing varieties and drought-tolerant crops and changing crop variety to match prevailing weather conditions, e.g., in Nigeria, Mozambique, Angola, and Kenya among other countries (Burnham and Ma 2016). The use of drought-tolerant maize has been documented in Angola, Benin, Kenya, and many other African countries (Nyasimi et al. 2014). Technological intervention in sub-Saharan Africa is observed to enhance sustainable intensification in post-harvest storage, choosing adapted crops and cultivars, intercropping and rotation with dual-purpose legumes, water harvesting, and irrigation to adapt local farmers to the adverse effects of climate change to agro-ecosystems. Similarly, technology is expected to enhance the choice for animal types and breeds that are better adapted to heat stress and dry conditions and to improve the storage of food and feed (Descheemaeker et al. 2016).

Moreover, technologies to reduce post-harvest losses are in consideration; e.g., the use of metal silo technology in household maize storage in Kenya has proved effective in protecting stored grains from attack by storage insect pests, sustaining maize supply while reducing the burden on the natural environment (Gitonga et al. 2013; Tefera et al. 2011). Other adaptation technologies promote water allocation efficiency for irrigation. For example, the water–energy–food (WEF) nexus is promoting the uptake of drip irrigation among smallholder farmers in Morrocco (He et al. 2006). WEF technology has been promoted as a demand-side management option for reducing water consumption while maintaining yields, particularly through minimizing nonproductive evaporative losses (Jobbins et al. 2015). In Ethiopia, other technologies that maximize precipitation utilization in dryland are promoted, e.g., increasing the length of the fallow period before planting to increase the amount of pre-plant stored water in the soil and retention of large amounts of crop residue on the soil surface to decrease runoff (Woyessa and Bennie 2007).

Climate Advisory and Extension

Climate advisories and information offer opportunities to inform farmers of climate-related risks, e.g., agrometeorological advisory program and weather and climate bulletins in Mali, and inform farmers on decisions such as variety selection, planting times, and the timing of inputs in Mali (Carr and Onzere 2018). Access to climate change information and extension services is encouraged in Kenya (Mutunga et al. 2017). However, 76% still have limited access to agricultural extension services (Kalungu and Leal Filho 2018). Extension services enhance access to information about new agricultural technologies and innovations. The use of weather information by farmers in Ghana is well documented (Belay et al. 2017). The use of early warning signs, weather forecasts, and agricultural extension services are indicated as important in enhancing the adaptation of farmers to climate variability in sub-Saharan Africa (Descheemaeker et al. 2016; Mbow et al. 2014). In East Africa, climate advisory and extension services remain low among smallholder farmers leading to limited capacity to use climate data (Atela et al. 2018; Singh et al. 2016) and climate illiteracy (Spires et al. 2014). However, the role of ICTs in scaling climate information services (CIS) is gaining attention (Atela et al. 2018). For example, short-term climate information through community worker initiative in Uganda provides 10-day, monthly, and seasonal weather forecasts to farmers on their mobile phones via SMS (Singh et al. 2016).

Access to weather information is observed to significantly influence the likelihood of adopting improved crop varieties, making adjustments in the timing of agricultural activities and investing in improved land management practices, and increasing fertilizer use among farmers in West Africa (Wood et al. 2014). Risk management practices that generate and disseminate agro-advisory services and weather information are important adaptive initiatives in Africa (Nyasimi et al. 2014).

Other Adaptive Initiatives

Increasing acreage as an initiative to enhance productivity is being practiced in some African countries, e.g., about 46.4% of the landmass in Mauritius is under agriculture, to enhance the production of crops and livestock (Luximon and Nowbuth 2010). The basis of increasing farm size is to increase food crop production to overcome low food crop productivity by maintaining normal productivity. However, this approach is facing sustainability concerns, e.g., continuous expansion of farmlands with a further clearing of the forest in Ghana has potential implications for deforestation, which will, in turn, exacerbate climate change in Ghana (Cobbinah and Anane 2016). Collective action involving collaboration among farmers is being practiced by local farmers and households. Community center approach to livestock management in Morocco (Kmoch et al. 2018), cooperatives and community-based development projects in South Africa (Belay et al. 2017), and social production and natural resource management-related groups in West Africa (Wood et al. 2014) have contributed both to limiting the risks and enhancing the capacity to adapt to climate change.

Change of pattern in planting times or moving animals is considered among some farmers in different areas: planting just before the onset of rains and moving animals, changing planting dates, and planting near a river in Kenya and Ghana (Belay et al. 2017; Mutunga et al. 2017; Descheemaeker et al. 2016); planting date adjustment in Ethiopia, Kenya, and Nigeria; planting early as possible after first rain (Belay et al. 2017; Oloo 2013), moving livestock to other areas, store fodder, and selling animals in South Africa (Belay et al. 2017); seasonal herd migration in sub-Saharan Africa (Descheemaeker et al. 2016); and temporary migration in Tanzania (Belay et al. 2017). *Capacity building* of climate change adaptation is recognized as a means of enhancing their ability to adopt desired initiatives related to technology adoption and diversification. Building capacity of farmers on climate change adaptation is documented in Mauritius (Luximon and Nowbuth 2010). In Mali, a farmer observer has been trained to enhance the capacity of most farmers in southern Mali (Carr and Onzere 2018). Capacity building of water user associations combined with local knowledge and scientific expertise is common in Morocco (Kmoch et al. 2018). In Zambia, the capacity of farmers is enhanced through conventional agricultural extension systems and participatory farmer interactions (Kuntashula et al. 2014). *Insurance* as a means of climate change adaptation is at trial in sub-Saharan Africa (SSA) to address climate-related risks faced by farmers (Descheemaeker et al. 2016). Weather-based index insurance is being tried in Mali (Carr and Onzere 2018). Adaptation planning in the form of insurance, with or without external support, is being tried in Kenya and Senegal (Crick et al. 2018). However, these products in general face low rates of adoption across SSA due to weakness of regulatory environment and financial facilities, basis risk, quality and availability of weather data, capacity building of stakeholders (farmer, insurer, and regulator), and lack of innovation for local adaptation and scalability (Ntukamazina et al. 2017).

The use of *credit facilitie*s has also been documented in Kenya (Mutunga et al. 2017) and Mauritius among other countries (Luximon and Nowbuth 2010). A study in Nigeria found a positive impact of commercial bank credits to food security by up to 8% (Osabohien et al. 2018). However, there is evidence that traditional credit use, among smallholder farmers, is extremely low in SSA (Adjognon et al. 2017), pointing to weak or no policy across many countries to enhance implementation.

Policy Initiatives for Climate Change Adaptation

Climate change is progressively hurting agricultural production in Africa (Kalungu and Leal Filho 2018; Crick et al. 2018; Mutunga et al. 2017; Belay et al. 2017; Burnham and Ma 2016). Adoption of suitable adaptation strategies is thus a pre-requisite to supporting the majority of smallholder farmers in reducing the effects of climate change (Hajer et al. 2018; Cobbinah and Anane 2016; Belay et al. 2017; Kmoch et al. 2018). Many factors, among them, are policies and markets and have been identified to define responses of the farmers to climate change shocks (Peter et al. 2017; Hummel 2016; Magnan et al. 2011; Kalame et al. 2009).

Policy to Enhance Diversification and Risk Management

Climate change adaptation policy in Mauritius made available 200 acres of agricultural land widening access to land for lease to small planters for agriculture diversification enhancing food production in the country (Luximon and Nowbuth 2010). Mauritius government advanced some incentives to farmers to support diversification, e.g., agricultural credit from the bank and subsidy on the price of farm input, guaranteed price for some farm produce, and offered some facilities to livestock breeders. Mauritius has the policy to limit food products locally produced to local consumption, while the country exploits the opportunity of cross-border initiatives (CBI) set by FAO with Madagascar, Mozambique, and Tanzania to increase production for domestic consumption (Luximon and Nowbuth 2010). A similar policy on price control also succeeded in Tunisia by reducing the exposure of Tunisian farmers to the food price volatility in the world markets and cautioning local households from the risk of scarcity in food supply (Chemingui et al. 2001). Participating in trade relations with the European Union (EU) and enhanced political cooperation has promoted the diversification of income sources from the trade of agricultural products (Chemingui et al. 2001).

Multi-sectoral Policies on Climate Advisory and Extension

In Namibia, the national-level development policy context of Vision 2030 guides Namibia's national, long-term development related to climate change adaptation. Climate change is mainstreamed in agriculture, disaster response strategies, and marine resources. The policy strengthens Namibia's meteorological services by supporting the work of the National Climate Change Committee in mainstreaming climate change adaptation and strengthening capacities to respond to it (Crawford and Terton 2016). The National Climate Change Strategy and Action Plan (NCCSAP) in Namibia has a direct link to enhancing climate change adaptation initiatives by setting up adaptation action on diversification of crops to adapt to erratic rainfall. It focuses on the increased use of improved crop varieties, conservation of indigenous livestock breeds, and diversification of livelihoods through tourism and wildlife conservancies (Crawford and Terton 2016). The policy aims to enhance capacities and synergies at individual, local, institutional, national, regional, and systematic levels to ensure successful implementation of climate change response activities and adequate funding resources for effective adaptation (Crawford and Terton 2016).

National-Level Policies for Adoption of Ecosystem-Based Initiatives

Policies on restoration or management of biodiversity and ecosystem services could confer multiple adaptive benefits, e.g., the use of shade trees in coffee farms (i.e., producing coffee as an agroforestry system), and could ensure the continued provision of key ecosystem services like pollination, natural pest control, conservation of water and soils, etc. as well as buffering coffee from extreme temperatures and rainfall, leading to more stable production under climate-related stresses (Vignola et al. 2015). Despite the growing interest in ecosystem-based adaptation, there has been little discussion of how this initiative could be enhanced among smallholder farmers while ensuring the continued provision of ecosystem services on which farming depends (Vignola et al. 2015). Policies on land use planning, forest-agriculture integration, and landscape-based adaptation are in line with enhancing local adoption of an ecosystem-based approach to climate change adaptation (Mbow et al. 2014). In South Africa, municipal climate adaptation plans use a sectoral approach to encourage greater interaction among different sectors and provided a clearer understanding of the needs and roles in climate adaptation (Roberts 2010). A specific case where policy implementation has been seen to influence the adaptation of technology and diversification initiatives was seen in the implementation of Global-GAP policy in the production of French beans in Kenya. The policy expected farmers to comply with certain climate adaptation strategies. The implementation supported changing crop variety, water harvesting, finding off-farm jobs, and soil conservation (Peter et al. 2017). The National Adaptation Plan 2015–2030 was developed by the Kenyan government to enhance adaptation capacity at macro-level. From the thematic perspective, Kenya's National Climate Change Action Plan (NCCAP 2018–2022) has prioritized adaptation initiatives focused on food security (GOK 2018).

Regional-Level Policies and Treaties

Regional treaties and international policies affect people's mobility in the context of climate change, and environmental change is strongly influenced by different international, regional, and national policies in the fields of migration, development, and environment. The crucial importance of mobility and migration within the West African region means that subregional initiatives, treaties, and regulations are particularly significant (Hummel 2016). Economic Community of West African States (ECOWAS)s common approach on migration enabling citizens to enter freely, reside, and settle in member countries is particularly important to livestock farmers in Senegal and Mali as one possible response to changing ecosystems (Hummel 2016).

Some regional commitments to enhance food security including the *Comprehensive Africa Agriculture Development Programme (CAADP)* and the subsequent *Malabo Declaration on Accelerated Agricultural Growth and Transformation for Shared Prosperity and Improved Livelihoods* are in place. However, the commitment to achieve the targets of the agreements may need upscaling. For example, the implementation of the agreement to allocate at least 10% of public expenditure to agriculture in Malabo Declaration remains low, with only Botswana and three other African countries including South Africa, Malawi, and Senegal engaging in public spending on agriculture worth above 10% (Mink 2016). Similarly, the implementation of the Paris Agreement to combat climate change by accelerating and intensifying climate actions remains weak with most countries making only small and cosmetic changes in transforming Intentional Nationally Determined Contributions (INDCs) to Nationally Determined Contributions (NDCs) (AFDB 2015).

National-Level Policies for Technology Adoption

Policies aligned to climate change adaptation could influence the adoption of technology. For example, Senegal Agriculture Programme was established in 2006 in response to an increase in outmigration from rural areas. This policy aimed at enhancing the use of technology in agriculture to generate sustainable incomes to attract young people to stay in their villages, enhancing rural development (Hummel 2016). Institutionalization of climate change adaptation into policy has the potential to enhance agricultural productivity. In Durban in South Africa, the climate adaptation policy ensuring that 50% of the food consumed by the rural poor is locally produced has led to the adoption of technology and diversification initiatives enhancing agricultural productivity (Roberts 2010). Policy context on climate change and agriculture in Ghana demonstrates a strong focus on the application of technology and innovation in agriculture as an engine for growth and development. Ghana's Shared Growth Development Agenda (GSGDA 2010–2013) highlighted the importance of facing climate change technological solutions such as drought-tolerant crop varieties and a transformation from rain-fed to irrigated agriculture, as well as reducing deforestation through agricultural expansion to improve the robustness of the sector to adapt to climate change (Sarpong and Anyidoho 2012). The government of Ghana has embarked on the implementation of a policy on reforestation through large-scale plantation development and small-scale on-farm regeneration activities to promote adaptation to climate change through ecosystem-based approaches. A similar initiative in reforestation has also been tried by the government of Burkina Faso (Kalame et al. 2009) and Kenya (Oloo 2013). However, there is fear that without proper planning, coordination, funding, and incentives to farmers for adoption, failure of reforestation policy could lead to large-scale deforestation and degradation, like in the case of Burkina Faso (Kalame et al. 2009).

Regional Policies and Trade-offs

In Morocco, food security trade-off policies are applied at the national level. Sufficiently transparent pricing model, subsidy, import substitution (increasing domestic production and reducing imports), intensification, and subsidizing farmers to grow cereals, either through deficiency payments or targeted input subsidies, have made cereal farming more lucrative (Magnan et al. 2011). Nonetheless, countries should observe that the application of trade-off policies does not contravene World Trade Organization regulations. Mauritius is often referred to as one of the very few developing countries to have overcome poverty and hunger by participating successfully in the globalization of the world market. Application of policy that enhances sector diversification to export-oriented agricultural diversification and industrialization aimed at reducing the dependence on food imports and increasing national food self-sufficiency. The implementation of the policy led to the attainment of national self-sufficiency in potatoes (Koop 2005). Policy on the modernization of agriculture and insulation of local farmers from external competition has permitted Tunisia to substantially increase its outputs, yields, and self-sufficiency rates in products considered as being strategic, such as cereals, vegetables, oil, and livestock products (Chemingui et al. 2001). Food self-sufficiency policy has influenced the development of irrigated farming in Botswana as a possible complement to rain-fed agriculture to enhance the production of cereals, mainly maize and sorghum (Lado 2001). The model food security pathway for African countries is summarized in Fig. 3.

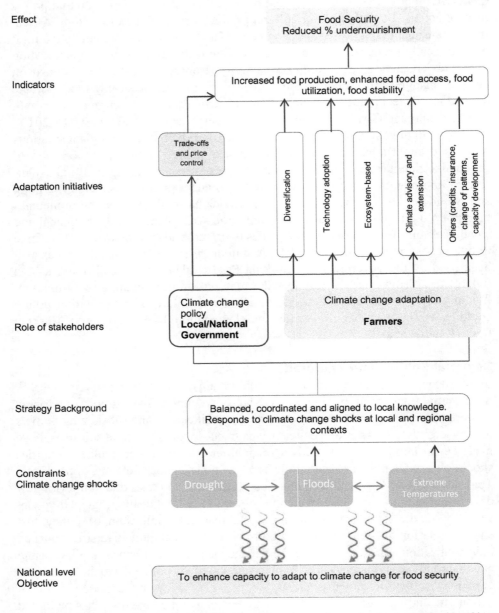

Fig. 3 A conceptual model of climate change adaptation pathway for food security in Africa. Balancing between climate change policy and adaptation (Authors own)

Conclusion

High prevalence of undernourishment was observed in six countries, all in SSA including Chad, Liberia, Central African Republic, the Democratic Republic of the Congo, Zambia, and Zimbabwe. Countries with a high land acreage under cereal production recorded reduced undernourishment as observed in Nigeria, Sudan, Ethiopia, Tanzania, and Niger. Five countries including Niger, Eritrea, Malawi, Swaziland, and Kenya experienced extreme climate variability in the last 10 years including droughts, floods, and extreme temperatures. Despite the extreme climate variability, Niger, Eritrea, Malawi, and Swaziland still registered average to high crop production index, an indication of effective climate change adaptation. Niger demonstrated effective adaptation for food security by registering the highest crop production index in extreme climate variability. However, Kenya appears to be the most predisposed by registering both high climate variability and below average crop production index.

Diversification and technology adoption were key strategies applied across countries. However, the uptake of technology by smallholder farmers is still low across many countries in SSA. There are efforts to mainstream climate change policy in agriculture at local, national, and regional levels to enhance technology adoption, capacity building, and funding. There is a need to strengthen policy to promote climate advisory and extension, climate literacy, and capacity to use ICTs to disseminate climate information services. A lack of commitment by governments to regional agreements to enhance food security and climate action has been observed. Only Botswana, South Africa, Malawi, and Senegal had spent on agriculture worth above 10% of national expenditure in line with the *Malabo Declaration*. Similarly, the implementation of the *Paris Agreement* to combat climate change by accelerating and intensifying climate actions remains to be weak in most countries.

The chapter maps the patterns of climate variability and undernourishment prevalence and establishes the linkages between policy initiatives and adaptation actions. Future academic explorations may focus on gaps in gender perspectives in climate change adaptation and effective ICTs for the dissemination of climate information services among smallholder farmers in SSA.

Acknowledgments Strathmore University provided information on the call for the research. Food and Agriculture Organization of the United Nation (FAO) through the online platform provided information on food security indicators and desktop data for analysis and World Development Indicator (WDI) for Environment and Climate Change data through the online data portal.

References

Abegunde VO, Sibanda M, Obi A (2019) The dynamics of climate change adaptation in sub-Saharan Africa: a review of climate-smart agriculture among small-scale farmers. Climate 7:1–23. https://doi.org/10.3390/cli7110132

Adjognon SG, Liverpool-Tasie LSO, Reardon TA (2017) Agricultural input credit in sub-Saharan Africa: telling myth from facts. Food Policy 67:93–105

AFDB (2015) Transitioning from INDCs to NDCs in Africa. AfDB CIF knowledge series, (11/2015). Africa Development Bank Group, Abidjan

AGRA (2014) Africa agriculture status report: climate change and smallholder agriculture in sub-Saharan Africa. AGRA, Nairobi

Atela J, Gannon KE, Crick F (2018) Climate change adaptation among female-led micro, small and medium enterprises in semi-arid areas: a case study from Kenya working paper. In: Leal Filho W (ed) Handbook of climate change resilience. Springer, Cham, pp 1–18. https://doi.org/10.1007/978-3-319-71025-9_97-1

Belay A, Recha JW, Woldeamanuel T, Morton JF (2017) Smallholder farmers' adaptation to climate change and determinants of their adaptation decisions in the Central Rift Valley of Ethiopia. Agric Food Secur 6:1–13. https://doi.org/10.1186/s40066-017-0100-1

Bongaarts J (2009) Human population growth and the demographic transition. Philos Trans R Soc B Biol Sci 364:2985–2990

Brandt M, Rasmussen K, Peñuelas J, Tian F, Schurgers G, Verger A, Mertz O, Palmer RBJ, Fensholt R (2017) Human population growth offsets climate-driven increase in woody vegetation in sub-Saharan Africa. Nat Ecol Evol 1:1–6

Burnham M, Ma Z (2016) Linking smallholder farmer climate change adaptation decisions to development. Clim Dev 8:289–311. https://doi.org/10.1080/17565529.2015.1067180

Carr ER, Onzere SN (2018) Really effective (for 15% of the men): lessons in understanding and addressing user needs in climate services from Mali. Clim Risk Manag 22:82–95. https://doi.org/10.1016/j.crm.2017.03.002

Cerdán CR, Rebolledo MC, Soto G, Rapidel B, Sinclair FL (2012) Local knowledge of impacts of tree cover on ecosystem services in smallholder coffee production systems. Agric Syst 110:119–130. https://doi.org/10.1016/j.agsy.2012.03.014

Chemingui M, Chemingui MA, Thabet C (2001) Internal and external reforms in agricultural policy in Tunisia and poverty in rural area Facilitating negotiations on the Arab Customs Union View project Trade structure and performance profile for Arab countries View project Internal and External Reform. https://www.researchgate.net/publication/229027151

Cobbinah PB, Anane GK (2016) Climate change adaptation in rural Ghana: indigenous perceptions and strategies. Clim Dev 8:169–178. https://doi.org/10.1080/17565529.2015.1034228

Conceição P, Levine S, Lipton M, Warren-Rodríguez A (2016) Toward a food secure future: ensuring food security for sustainable human development in sub-Saharan Africa. Food Policy 60:1–9. https://doi.org/10.1016/j.foodpol.2016.02.003

Crawford A, Terton A (2016) Review of current and planned adaptation action in Namibia. International Development Research Centre. http://www.idrc.ca/cariaa

Crick F, Eskander SMSU, Fankhauser S, Diop M (2018) How do African SMEs respond to climate risks? Evidence from Kenya and Senegal. World Dev 108:157–168. https://doi.org/10.1016/j.worlddev.2018.03.015

Dasgupta P, Morton JF, Dodman D, Karapinar B, Meza F, Rivera-Ferre MG, Toure Sarr A, Vincent KE (2014) Rural areas. In: Field CB, Barros VR, Dokken DJ, Mach KJ, Mastrandrea MD, Bilir TE, Chatterjee M, Ebi KL, Estrada YO, Genova RC, Girma B, Kissel ES, Levy AN, MacCracken S, Mastrandrea PR, White LL (eds) Climate change 2014: impacts, adaptation, and vulnerability. Part A: global and sectoral aspects. Contribution of working group II to the fifth assessment report of the Intergovernmental Panel on Climate Change. Cambridge University Press, Cambridge, pp 613–657

Descheemaeker K, Oosting SJ, Homann-Kee Tui S, Masikati P, Falconnier GN, Giller KE (2016) Climate change adaptation and mitigation in smallholder crop–livestock systems in sub-Saharan Africa: a call for integrated impact assessments. Reg Environ Chang 16:2331–2343. https://doi.org/10.1007/s10113-016-0957-8

Gitonga ZM, De Groote H, Kassie M, Tefera T (2013) Impact of metal silos on households' maize storage, storage losses and food security: an application of a propensity score matching. Food Policy 43:44–55

GOK (2018) Republic of Kenya Ministry of Environment and Forestry National Climate Change action plan 2018–2022, vol I. Ministry of Environment and Forestry, Nairobi

Hajer A, Mohsen BS, Hatem A, Hichem K, Mahouachi M, Beckers Y, Hammami H (2018) Climate change-related risks and adaptation strategies as perceived in dairy cattle farming systems in Tunisia. Clim Risk Manag 20:38–49. https://doi.org/10.1016/j.crm.2018.03.004

He L, Tyner WE, Doukkali R, Siam G (2006) Policy options to improve water allocation efficiency: analysis on Egypt and Morocco. Water Int 31:320–337

Hummel D (2016) Climate change, land degradation and migration in Mali and Senegal – some policy implications. Migr Dev 5:211–233. https://doi.org/10.1080/21632324.2015.1022972

IPCC (2007) Climate change 2007: synthesis report. In: Pachauri RK, Reisinger A (eds) Contribution of working groups I, II and III to the fourth assessment report of the intergovernmental panel on climate change [core writing team]. IPCC, Geneva, Switzerland, p 104

IPCC (2013) Climate change 2013: the physical sciences basis. IPCC, Geneva, Switzerland

IPCC (2018) Global warming of 1.5 oC. In IPCC – Sr15. https://report.ipcc.ch/sr15/pdf/sr15_spm_final.pdf%0A http://www.ipcc.ch/report/sr15/

Jobbins G, Kalpakian J, Chriyaa A, Legrouri A, El Mzouri EH (2015) To what end? Drip irrigation and the water-energy-food nexus in Morocco. Int J Water Res Dev 31:393–406

Kalame FB, Nkem J, Idinoba M, Kanninen M (2009) Matching national forest policies and management practices for climate change adaptation in Burkina Faso and Ghana. Mitig Adapt Strateg Glob Chang 14:135–151. https://doi.org/10.1007/s11027-008-9155-4

Kalungu JW, Leal Filho W (2018) Adoption of appropriate technologies among smallholder farmers in Kenya. Clim Dev 10:84–96. https://doi.org/10.1080/17565529.2016.1182889

Keil A, Zeller M, Franzel S (2005) Improved tree fallows in smallholder maize production in Zambia: do initial testers adopt the technology? Agrofor Syst 64:225–236

Khacheba R, Cherfaoui M, Hartani T, Drouiche N (2018) The nexus approach to water–energy–food security: an option for adaptation to climate change in Algeria. Desalin Water Treat 131:30–33. https://doi.org/10.5004/dwt.2018.22950

Khaldi R, Mohamed J, Khaldi G (2012) Impacts of climate change on the small ruminants farming systems in northwestern Tunisia and adaptation tools. In: New approaches for grassland research in a context of climate and socio-economic changes. CIHEAM, Zaragoza, pp 427–431

Kimaro AA, Timmer VR, Chamshama SAO, Mugasha AG, Kimaro DA (2008) Differential response to tree fallows in rotational woodlot systems in semi-arid Tanzania: post-fallow maize yield, nutrient uptake, and soil nutrients. Agric Ecosyst Environ 125:73–83

Kiptot E, Franzel S, Degrande A (2014) Gender, agroforestry and food security in Africa. Curr Opin Environ Sustain 6:104–109. https://doi.org/10.1016/j.cosust.2013.10.019

Kmoch L, Pagella T, Palm M, Sinclair F (2018) Using local agroecological knowledge in climate change adaptation: a study of tree-based options in northern Morocco. Sustainability (Switzerland) 10:3719. https://doi.org/10.3390/su10103719

Koop K (2005) Food security in the era of globalisation-the case of Mauritius 49 governance and food security in the era of globalization: the case of Mauritius, vol II. https://halshs.archives-ouvertes.fr/halshs-00265114

Kuntashula E, Chabala LM, Mulenga BP (2014) Impact of minimum tillage and crop rotation as climate change adaptation strategies on farmer welfare in smallholder farming systems of Zambia. J Sustain Dev 7:95–110. https://doi.org/10.5539/jsd.v7n4p95

Lado C (2001) Environmental and socio-economic factors behind food security policy strategies in Botswana. Dev South Afr 18:141–168. https://doi.org/10.1080/037/68350120041875

Lasco RD, Delfino RJP, Catacutan DC, Simelton ES, Wilson DM (2014) Climate risk adaptation by smallholder farmers: the roles of trees and agroforestry. Curr Opin Environ Sustain 6:83–88. https://doi.org/10.1016/j.cosust.2013.11.013

Luximon Y, Nowbuth MD (2010) A status of food security in Mauritius in face of climate change. Eur Water 32:3–14

Magnan N, Lybbert TJ, McCalla AF, Lampietti JA (2011) Modeling the limitations and implicit costs of cereal self-sufficiency: the case of Morocco. Food Secur 3:49–60. https://doi.org/10.1007/s12571-010-0103-2

Mbow C, Van Noordwijk M, Luedeling E, Neufeldt H, Minang PA, Kowero G (2014) Agroforestry solutions to address food security and climate change challenges in Africa. Curr Opin Environ Sustain 6:61–67. https://doi.org/10.1016/j.cosust.2013.10.014

Mink SD (2016) Findings across agricultural public expenditure reviews in African countries, vol 1522. International Food Policy Research Institute, Washington, DC

Mutunga EJ, Ndungu CK, Muendo P (2017) Smallholder farmers perceptions and adaptations to climate change and variability in Kitui County, Kenya. J Earth Sci Clim Change 8(3). https://doi.org/10.4172/2157-7617.1000389

Namirembe S, Leimona B, Van Noordwijk M, Bernard F, Bacwayo KE (2014) Co-investment paradigms as alternatives to payments for tree-based ecosystem services in Africa. Curr Opin Environ Sustain 6:89–97. https://doi.org/10.1016/j.cosust.2013.10.016

Ntukamazina N, Onwonga RN, Sommer R, Rubyogo JC, Mukankusi CM, Mburu J, Kariuki R (2017) Index-based agricultural insurance products: challenges, opportunities and prospects for uptake in sub-Sahara Africa. J Agric Rural Dev Trop Subtrop 118(2):171–185

Nyasimi M, Amwata D, Hove L, Kinyangi J, Wamukoya G (2014) Evidence of impact: climate-smart agriculture in Africa. http://www.ccafs.cgiar.org

Oloo G (2013) Evaluation of climate change adaptation strategies and their effect on food production among smallholder farmers in Bungoma County, Kenya. Egerton

Osabohien R, Afolabi A, Godwin A (2018) An econometric analysis of food security and agricultural credit facilities in Nigeria. Open Agric J 12(1):227–239

Peter SO, Chris AO, John M, Rose AN (2017) Effect of global-GAP policy on smallholder French beans farmers climate change adaptation strategies in Kenya. Afr J Agric Res 12:577–587. https://doi.org/10.5897/ajar2017.12149

R Core Team (2020) R: a language and environment for statistical computing. R Foundation for Statistical Computing, Vienna, Austria

Roberts D (2010) Prioritizing climate change adaptation and local level resilience in Durban, South Africa. Environ Urban 22:397–413. https://doi.org/10.1177/0956247810379948

Sarpong DB, Anyidoho NA (2012) Climate change and agricultural policy processes in Ghana. http://www.future-agricultures.org

Serdeczny O, Adams S, Baarsch F, Coumou D, Robinson A, Hare W, Schaeffer M, Perrette M, Reinhardt J (2016) Climate change impacts in sub-Saharan Africa: from physical changes to their social repercussions. Reg Environ Chang 17:1585–1600

Silvestri S, Sabine D, Patti K, Wiebke F, Maren R, Ianetta M, Carlos QF, Mario H, Anthony N, Nicolas N, Joash M, Lieven C, Rufino MC (2015) Households and food security: lessons from food secure households in East Africa. Agric Food Secur 4:23. https://doi.org/10.1186/s40066-015-0042-4

Singh C, Urquhart P, Kituyi E (2016) From pilots to systems: barriers and enablers to scaling up the use of climate information services in smallholder farming communities. CARIAA working paper no. 3. International Development Research Centre, Ottawa/London. http://www.idrc.ca/cariaa

Spires M, Shackleton S, Cundill G (2014) Barriers to implementing planned community-based adaptation in developing countries: a systematic literature review. Clim Dev 6:277–287. https://doi.org/10.1080/17565529.2014.886995

Tefera T, Kanampiu F, De Groote H, Hellin J, Mugo S, Kimenju S, Beyene Y, Boddupalli PM, Shiferaw B, Banziger M (2011) The metal silo: an effective grain storage technology for reducing post-harvest insect and pathogen losses in maize while improving smallholder farmers' food security in developing countries. Crop Prot 30:240–245

UNFCC (2006) Technologies for adaptation to climate change. Climate Change Secretariat (UNFCCC), Bonn, Germany

Vignola R, Harvey CA, Bautista-Solis P, Avelino J, Rapidel B, Donatti C, Martinez R (2015) Ecosystem-based adaptation for smallholder farmers: definitions, opportunities and constraints. Agric Ecosyst Environ 211:126–132. https://doi.org/10.1016/j.agee.2015.05.013

Wood SA, Jina AS, Jain M, Kristjanson P, DeFries RS (2014) Smallholder farmer cropping decisions related to climate variability across multiple regions. Glob Environ Chang 25:163–172. https://doi.org/10.1016/j.gloenvcha.2013.12.011

Woyessa YE, Bennie ATP (2007) Tillage-crop residue management and rainfall-runoff relationships for the Alemaya catchment in Eastern Ethiopia. Sh Afr J Plant Soil 24:8–15

6

Managing Current Climate Variability can Ensure Water Security Under Climate Change

Mike Muller

Contents

M. Muller (✉)
Wits School of Governance, University of the Witwatersrand, Johannesburg, South Africa
e-mail: mike.muller@wits.ac.za

Abstract

Water resources will be significantly impacted upon by climate change, and these impacts will be transmitted to the many sectors and services dependent on them. The nature, extent, and timing of these impacts remain uncertain, but the long lifetime of water infrastructures requires that their planning, development, and operations should be resilient to climate changes. An effective approach is to focus on the management of current climate variability as it relates to water, which strengthens the ability of communities and countries to foresee, manage, and adapt to the impacts of longer-term climate change on water-related activities. This approach is illustrated by cases from Southern and Eastern Africa.

Current "stationary" stochastic methods of hydrological analysis can still be used under assumptions of a "dynamic stationarity" although more regular updating of hydrological data will be required. Methodologies to evaluate economic dimensions of risk reduction introduce additional uncertainties but may help decision-makers to understand the risks and opportunities. Diversification of sources and sequencing of resource development pathways are helpful strategies to adapt to climate change but must ensure that risks affecting different sources are not correlated. Attention must also be given to demand-side interventions in order to reconcile supply and demand, and these perspectives must be shared with social, economic, and political actors to ensure that strategies are communicated, understood, and supported by the wider community.

Keywords

Water resources · Climate adaptation · Climate variability and change · Water supply · Hydropower · Hydrology · Stationarity · Utility economics · Public finance

Introduction

There is international agreement at the UNFCCC about the need for collective action to address the likely impacts of global warming on human societies. One response has been to encourage all sectors of society to identify and implement actions that can help adapt to the emerging impacts of climate change and work to mitigate its drivers.

Water resources and the services dependent on them are an integral part of the climate system (Chahine 1992). Because of this, the managers of water resource systems and of the services that depend on them must develop appropriate responses to the potential impacts of climate change (Strzepek et al. 2011).The African continent is considered to be particularly vulnerable to these impacts both because of their magnitude and because African societies have less physical, financial, and human resources available to address these impacts.

A further challenge is that the impacts of climate change on water resources are local, diverse, and not well-characterized. Unlike the general global warming trend, for which there is robust evidence that is consistent with the predicted impact of anthropogenic activities, direct evidence of significant changes in hydrological variables is not nearly as strong and consistent.

For example, it appears that at a global level, precipitation is increasing (Adler et al. 2017). However, models predict increases in drought frequency over significant areas due to warming (Ukkola et al. 2020) although empirical evidence in support of this is limited (Hegerl et al. 2019) and predictions are based on various definitions of drought.

While warming trends are a direct result of anthropogenically induced atmospheric changes that act through a single dominant mechanism at global level, the impacts on the main hydrological processes – evaporation, precipitation, runoff, and infiltration – are secondary and tertiary effects that are further influenced by a wide variety of mechanisms at a local, regional, and global level. So while there is only a limited increase in global precipitation averages, there are more extensive regional variations and substantial changes in parameters such as rainfall intensity (Trenberth 2011) which, coupled with the impact of warmer temperatures on aridity, lead to changes in tertiary variables such as stream flow.

The likely impact of climate change on hydrological processes is generally still too uncertain to make it possible for water resource planners and managers to identify locally specific adaptive measures. So climate change models and predictions cannot usually guide well-founded planning and operational decisions although they do help to suggest the boundaries for extreme events.

However, policy-makers like the UN High-Level Panel on Water want to know what impacts global warming will have on water systems and how they will be dealt with. It is suggested that an appropriate response is to continue to build on the great strength of best practice in water resource science and management, which is its ability to characterize and manage climatic *variability*. This requires that the approaches used move beyond the assumptions of a stationary hydrology and can accommodate a changing climate (Brekke et al. 2009).

Yet water managers are already responding to climate change through their management of current climate variability. However, to make this case convincingly, they must demonstrate that current methodologies are adequate for the task. To address this in an African context, this chapter:

- Characterizes the "difficult hydrologies" that pose particular challenges in Africa and shows how climate change science has helped to understand the climate variability inherent in those hydrologies
- Considers some analytical tools that are used to manage under assumptions of "stationary hydrology" and whether they can be adapted reflect "dynamic stationarity"

– Presents six African cases that illustrate how the management of current variability is supporting adaptation to potential climate change impacts

Drawing on lessons from these cases, some conclusions are drawn about approaches that, while deriving from current hydrologies, are more explicitly adaptive and will be resilient in an uncertain future.

A supply-side focus is taken because the availability of water is most directly impacted upon by climate change. The demand side is equally important and challenging, but human behavior and social institutions that guide it merit separate consideration. The reconciliation of supply and demand remains the ultimate technical goal of water management.

Characterizing Hydrological Variability and Climate Change Impacts

Current Hydrological Variability

Climate warming due to anthropogenic activity is a global phenomenon, but its specific impacts on water resources depend on local conditions, including current climatic variability. Water resource managers need to characterize that current variability if they are to manage it successfully. This characterization provides the foundation from which to review the impacts of climate change on water resources and the hydrological cycle.

The primary driver of the hydrological cycle is the interaction between incoming solar energy and the large mass of water in the oceans. However, water managers necessarily focus on rainfall, whose duration, intensity, and distribution in space and time drive other elements of the hydrological cycle: surface runoff, infiltration, and evapotranspiration. The interaction between these variables then determines the stream and groundwater flows, whose magnitude and fluctuation are the "raw material" of water resource management.

"Easy" Versus "Difficult" Hydrologies

From this perspective, climate creates the hydrological context both for the "natural" biosphere and for human communities and their activities that depend on it. But it is the variability of climate as much as the average values of its component elements that determines whether the resulting hydrological context enables communities to establish and sustain productive and congenial environmental niches. Grey and Sadoff (2007) echo previous investigators, suggesting that climatic variability has been a determining factor in the social and economic development of different societies.

A nation's hydrology will clearly affect the level of institutions and investment required to achieve water security. The absolute levels of water resource availability, its inter- and intra-annual variability and its spatial distribution, coupled with the demand for water, will largely determine the institutions and the types and scale of infrastructure needed to manage, store and move the resource. The resilience of the structure of the economy to water shocks, together with societal resilience and risk aversion will also be determinants of the level of investment required for specific countries to reach the tipping point of water security.

A distinction is drawn between "easy" and "difficult" hydrological legacies. An "easy" hydrology is characterized by adequate rainfall with little variation between seasons and years. Such hydrologies sustain predictable perennial river flows, supported by reliable groundwater availability. The predictability of the hydrology and its relative consistency makes the management of water for different purposes relatively easy, facilitating the conceptualization and operation of infrastructure as well as the establishment of the rules that govern the entitlements and obligations of water users.

Regions with difficult hydrologies require stronger institutions as well as higher levels of physical investments in order to support basic activities such as agriculture or even simply to meet domestic needs. One reason that many poor countries remain poor is that they could not accumulate the resources needed to achieve the water security on which higher levels of development depend. As summarized, difficult hydrologies include:

- Absolute water scarcity
- Low-lying lands subject to severe flood risk
- Markedly seasonal rainfall which requires the storage of water
- High inter-annual climate variability, where unpredictable risks require over-year storage
- A combination of extreme intra-annual seasonality and inter-annual variability

This crude distinction between difficult and easy hydrologies highlights the need to understand existing climate variability before focusing on the potential impacts of climate change. At the least, water managers need to understand the key systems that drive their local climate and the extent of uncertainty about the resulting rainfall patterns and related hydrological parameters in order to identify sustainable strategies for their water resource development and management.

Climate Change Science Has Helped to Characterize Climate Variability

Paradoxically, climate change science must build on an understanding of "normal" climate variability in order to attribute any phenomenon to climate change rather than simply a "normal" extreme. So while the Sahel droughts of the 1970s/1980s were often attributed to climate change, evidence of subsequent "rainfall recovery" suggests that this was simply long-cycle climate variability, perhaps aggravated by land- and water-use changes (Sidibe et al. 2018).

There are similar debates about changes in the intensity and frequency of the occasional and unpredictable tropical storms, an important factor in the hydrology of Southeastern Africa (Malherbe et al. 2012). Claims that those storms were becoming more frequent and intense due to global sea warming have subsequently been challenged, since the 35-year-long satellite data set on which they were based is relatively short; it is now suggested that tropical storm formation was mostly influenced by long-cycle variations in the phenomena such as ENSO and the Indian Ocean Dipole (Chan 2006).

The hypothesis that storm frequency is increasing due to global warming is also not supported by later studies, and it is even suggested that south of the tropic of Capricorn, where storm numbers were expected to increase, they have actually decreased (Pillay and Fitchett 2019). However, the debate continues, and there is continuing uncertainty about likely changes in cyclone frequency and their regional prevalence (Knutson et al. 2020).

Variations in lake levels are also often attributed to climate change with recent fluctuations in Africa's lakes Victoria, Chad, and others specifically cited, although with acknowledgment that human influence is also a factor. Yet many of Africa's shallower African lakes are known for their variable levels (Conway 2005).

Lake levels are arguably more useful as indicators of the impact of climate *variability* than climate *change* since they are determined by the climate over the lakes and their catchment areas as well as the impact of human activity. A recent global review found that background climate variation due to multi-decadal climatic oscillations such as ENSO accounted for 58% of the variation in lake levels; a further 10% variation was due to normal seasonal effects. It was concluded that apparent trends attributed to anthropogenic activities were often exaggerated by this "normal" background variation and that due attention should be given to background climate variation before claiming climate change impacts (Kraemer et al. 2020).

These examples show that it is necessary to correct the current popular (and academic) discourse in Africa and elsewhere which sees water managers, policy-makers and political heads attribute many extreme events to climate change, although climate records show that neither the event nor its frequency of occurrence is inconsistent with patterns evident in the historical record.

Predictions of Climate Change Impacts on Hydrological Processes

There is no doubt that global warming is already having a variety of impacts on hydrological systems. However, while some impacts are known, others are uncertain and are likely to vary significantly from place to place. These uncertainties need to be better understood and communicated by practitioners to policy-makers and publics.

Some climate change impacts are directly related to the impact of anthropogenic warming on specific hydrological variables. While the water-holding capacity of air increases with temperature, this does not necessarily mean that global rainfall will increase at the same rate. However, recent work concludes that observations confirm modeling predictions that rainfall does increase with rising temperatures and that this is associated with a change in the intensity and timing of rainfall events (Giorgi et al. 2019). It is also evident that there are regional changes in the distribution of rainfall although uncertainties remain about these important relationships (Herold et al. 2017).

Similarly uncertain is the impact of warming climates on soil moisture, a hydrological variable that underpins the agricultural activities on which human societies depend and which are, by far, the largest direct "users" of water. Aridity, the ratio between precipitation and potential evapotranspiration, is expected to impact on surface water runoff, a critical process for water management.

Aridity might be expected to increase with global warming since evapotranspiration is greater under higher temperatures, but this is not universal. Globally, some arid areas (in the Americas) have become wetter, whereas previously semi-humid areas (particularly in Asia) became drier. And significant discrepancies are reported between model predictions and direct observations leading to warnings that the general conclusion that climate change will result in "overall drier conditions across the globe" might be "at least partly misleading" (Greve et al. 2019).

The discrepancies are attributed, in part, to the fact that the increased CO_2 that drives global warming also changes the vegetation types, which may use water more efficiently. As the authors explain, "....changes in atmospheric CO_2 break the existing correlation between hydrology and ecology by changing the water use efficiency of photosynthesis."

These issues are important for sub-Saharan Africa where bush encroachment is reducing the animal carrying capacity of semiarid rangelands although human changes in land use are at least as important as climate change in driving this process (Venter et al. 2018). What is important for water managers is that changes in aridity and land cover will change the relationship between rainfall and runoff, potentially affecting water resource availability (Zareian et al. 2017) and leading to significant stream flow reductions.

These processes are complex, and outcomes will be determined by locally specific conditions. The proportion of rainfall that reaches a stream or recharges groundwater is determined by the amount of rainfall, its intensity and duration, prior soil moisture, as well as temperature and land cover. One response to this complexity has been to apply catchment models, developed to provide guidance on rainfall-runoff relationships. But these models are based on historical circumstances that may not be appropriate under new conditions of climate and land cover. In Australia, historical rainfall-runoff models might produce valid results under future climates, but only if rainfall changes were relatively modest (Vaze et al. 2010). In Europe while runoff correlates closely with soil moisture, relations with precipitation and temperature are weaker, and a recent survey identified a better understanding of runoff dynamics as a priority "unsolved question" for hydrologists (Blochl et al. 2019).

Tools to Manage "Dynamic Stationarity"

There will clearly be continuing uncertainty about climate change-driven hydrological trends due to the wide diversity of water resource contexts in which hydrological processes occur. Furthermore, substantial "normal" variability will often mask smaller climate change impacts.

It is thus appropriate for practitioners to focus explicitly on managing climate variability. This is not an argument for "stationarity" (the assumption that climate is not changing) nor a suggestion that water resource planners should ignore warming-related climate change. Rather, it is a practical response. While it has long been accepted that hydrological analysis must reflect the changing climate (Moss and Tasker 1987), tools developed under assumptions of "stationary" hydrology may yet be used under the new conditions of hydrological dynamism (Milly et al. 2008).

Right Tools for the Right Job: But which Job?

Water resource practitioners use a wide variety of tools to guide their decisions, but their tools' limitations and their adaptation to address changing climates must be considered in the context of the functions that they support. These include:

– Monitoring and information management, about both resource availability and resource use
– Planning, to determine trends in resource availability and use and to inform options, identification, and analysis
– Allocation, to ensure that water use achieves society's objectives within the constraints imposed by the resource itself
– Development of infrastructure, to make water available for use as well as to maintain its quality
– Sustainable operation of complex infrastructure systems
– Protection of the environmental condition of the resource

The timeframe and nature of guidance that will be required vary dramatically between the functions, and the tools provided have to be appropriate for the purpose:

– Planning requires projections of the yield available from different sources and systems, often over very long time horizons (50–100 years in the case of large systems) with indications of assurance derived from assumptions about the variability of the resource. While climate change is clearly relevant over long time horizons, planning does not necessarily commit to specific actions until firm decisions are taken.
– Infrastructure must optimize output based on design requirements and be guided by estimates of the magnitudes of extreme events that could lead to failure. The long lifetime of water resource infrastructure limits the subsequent adoption of alternative options and enforces a degree of "path dependence" (Hüttl et al. 2016).

– System operation needs short-term (seasonal up to 5 years) guidance to maintain
  supply assurances, while flood routing requires real-time models.
– Water allocations to users must be informed by "normal" availability of supply
  but allow adjustment for contingencies such as drought as well as to cater for
  future climate change (which is often missing from legal frameworks despite
  earlier warnings (Trelease 1977)).

In summary, water resource managers must be enabled to (i) make reasonable
predictions about the potential yield of sources and systems over time; (ii) provide
guidance based on variability parameters to support system restrictions during
drought; (iii) provide robust, conservative information about risks posed by extreme
flood events; and (iv) operate their systems to achieve reasonably predictable outputs
under variable conditions.

What Tools Are Available?

Simple rainfall-runoff models, developed to support urban drainage design and flood
estimation, were of limited value in larger catchments, particularly in the absence of
information on rainfall characteristics. The development of stochastic techniques to
provide rainfall and river flow estimates and guide reservoir operation greatly
expanded the analytical capabilities of water planners.

The complex relationships between hydrological variables, the diversity of situations, and the volume of data involved meant that key concepts could only be
translated into tools for practitioners when modern computing power became available. The Harvard Water Program developed many methodologies using "synthetic"
or stochastic hydrologies (Maass et al. 1962) not just to solve hydrological questions
but also to support the economic optimization of water projects.

There is now a wide variety of hydrological tools and methodologies that have
been applied. Loucks and van Beek's extensive compendium (2017) (Loucks and
Van Beek 2017) links the technicalities of methodologies and models to their
application in real-world planning problems. These are supported by new streams
of data from remote sensing although this does not fully compensate for the decay in
the physical observation network (World Bank 2018a). The challenges posed by
climate change have led to suggestions that a new Harvard Water Program is needed.

To address some of the functions outlined above, tools used include the
following:

Stochastic Methodologies to Estimate Flow and Rainfall

The stochastic methodologies described above have transformed water management.
When applied with the requisite caution and skill, they enable planners to extend
short flow or rainfall records to provide reasonable estimates of mean and extreme
flows at different levels of assurance as well as to provide synthetic "records" for
ungauged rivers and catchments. Stochastic models lend themselves to testing
hypotheses about the potential impacts of changes in rainfall patterns and (given

the computing power now available) can generate and test large numbers of synthetic sets of rainfall and flow to determine sensitivities to different climatic circumstances. Climate change does, however, mean that the value of such methodologies depends on the regular updating of the data on which they are based.

Catchment-Scale Rainfall-Runoff Models to Support Yield and Operations

Catchment-based rainfall-runoff models are also widely used to estimate the potential yield of different sources and systems and to support operations ranging from drought management to flood routing. The rainfall-runoff relationships can be extrapolated from known areas to similar, poorly gauged catchments to predict stream flows and guide water management more generally (Vogel 2017). Such models can provide a structured basis for yield estimation taking account of new flow and rainfall records and changes in land use (see, for instance, the South African framework which has evolved over the past 60 years) (Bailey and Pitman 2016). While the maintenance of these models requires consistent investment in both physical data collection and subsequent processing, they provide an invaluable basis for the planning, expansion, and operation of large systems. The evolution of the planning and operational models for South Africa's Vaal River System (see below) provides an instructive insight.

Methods to Determine the Scale of Maximum Probable Events

A particular challenge in designing large water resource infrastructure is to ensure that it is resilient to the most extreme event "likely." The primary determinant of the scale of the "probable maximum flood" is the probable maximum precipitation. Since the objective in determining the PMF is to identify a worst-case scenario, this is a case where the application of multiple climate models can be useful since there will be a strong argument for selecting the highest value generated (Gangrade et al. 2018). For these kinds of events, in large catchments, catchment dynamics and land-use changes have less influence.

Right Tools in the Right Place: At the Right Time

One consequence of the huge diversity in the local hydrological regimes on which societies' water supplies depend is that technical paradigms that work well in one context have often been promoted inappropriately in others (Woodhouse and Muller 2017). Aid-dependent African countries have been particularly vulnerable to this trend because approaches promoted often reflected donor country conditions or the educational system in rich countries simply assumes that its approach is globally appropriate (Briscoe 2010). The promotion of a reliance on "green infrastructure" rather than built infrastructure is a recent incarnation of this trend (Muller et al. 2015).

While environmentally focused approaches may be appropriate in temperate developed countries, they do not help communities under more challenging climates in Africa, Asia, and Latin America where populations and economies are growing rapidly. Aside from the socioeconomic conditions, there are dramatically different topographies, geologies, and climatic conditions of aridity and often greater inter-seasonal and inter-annual variability of temperature and precipitation.

There are stark contrasts in the physical context for water resource management even in developed regions, between Europe's temperate northern regions and its Mediterranean south and temperate northeastern USA, its humid, subtropical south-east, giving way to semiarid southwest. In each region, the character of water resources and thus the options available to address water resource challenges are different, and different tools of analysis will be used. A further key differentiator is the human and financial resources that can be brought to bear: "With increasingly 'difficult' hydrology, the level of institutional refinement and infrastructure invest-ment needed to achieve basic water security becomes significantly greater than in temperate (and less variable) climates" (Grey and Sadoff 2007).

There must be appropriate responses to current variability and uncertainty before the particular challenges of climate change can be addressed. For many countries, the priority problems relate to present water insecurity, not future climate change. As the cases below show, responses to, for instance, urban water shortages are unlikely to be much altered by considerations of climate change.

When solutions proposed could increase future costs and risks, climate change considerations become important. What is most immediately important in (rela-tively) poor communities and countries that do not have acceptable levels of water security is the efficient and effective use of scarce resources to deal with current variability. Many recent water "crises" popularly attributed to climate change have been the consequence of a failure to prepare for current variability. Opposition to long-run investment decisions (Matalas 1997) aggravates immediate problems, with limited evidence of long-run negative consequences. Often, it reflects policy prefer-ences rather than any well-founded evidence of harm or risk (Muller et al. 2015)

From Tools to their Application: Some Case Studies

Water management practitioners have available a plethora of tools of increasing complexity and cost (as measured in human resources, finances, and data) to help them to manage climate variability. Yet the extent to which these tools can help water managers to address the challenges of climate change is limited. How should the practitioners proceed?

This is more than a technical issue. Water is managed to achieve a wide variety of societal goals, and water managers operate under a range of technical, financial, as well as social and political constraints and produce a variety of economic benefits. Tools to analyze the performance of water projects are also needed to assess their financial risks and economic returns in many different contexts.

A new mining or agricultural business may require a secure water supply. Governments seek to reduce poverty and promote greater economic inclusion by providing irrigation water, enabling poor farming communities to improve their productivity. Power utilities seek competitive, reliable energy sources. Urban managers, concerned about the disruptive effects of water restrictions due to drought, want to enhance water security.

Water managers and their institutions must be able to respond authoritatively to both political decision-makers and their wider communities. Most stakeholders will be concerned about the risks that climate change is assumed to pose and will need some assurances that these have been adequately addressed.

In this section, a number of cases from Southern and Eastern Africa are presented that show how water managers have used available tools to achieve specific water management objectives by addressing climate variability and concerns about climate change. The cases locate climate variability and change challenges in a wider socioeconomic context and consider whether the approaches adopted have increased resilience to potential climate change impacts. Three of the cases were included in a World Bank study of resilience in African infrastructure which aimed to develop processes that support "robust decision-making" (RDM) (Cervigni et al. 2015).

Cape Town's Day Zero Drought: A Failure to Acknowledge Risks of Climate Variability

Between 2015 and 2018, droughts in the extreme south west of South Africa saw the City of Cape Town suffering serious water restrictions. It was widely reported that the city faced a "Day Zero" on which its water supplies would "run out." These events, copiously documented, were widely attributed to climate change.

However, over the previous decade, there had been repeated recommendations from national technical and planning agencies calling for investments to augment, by 2015, the supply capacity of the Western Cape Water Supply System (WCWSS), on which the city depends (Muller 2018). These recommendations were based on a long-term WCWSS strategy study, informed by a hydrologically based system model that had successfully been used for two decades.

The model's projections of the supplies reliably available from existing sources were set against projections of water demand in a "reconciliation" process. This identifies any requirement for increased supply, the primary driver being the needs of Cape Town's growing population. The interventions proposed were sequenced on a "least-unit-cost" basis. Records of meetings in 2013 and 2014 suggest that the city believed that its efforts to manage demand had constrained growth and that further supply expansion would only be required in 2022.

Even before the drought ended, implementation had begun of infrastructure projects that had earlier been rejected – including water reuse and groundwater development as well as additional surface water. Consideration is also being given to a large desalination installation which could be used during drought periods.

The vulnerabilities of the WCWSS are now obvious. The system depends on rainfall over a small (800km^2) area of mountain catchments supplying dams whose storage is less than 2 years of "normal" use. Climate change dynamics are also under review. Cape Town lies at the edge of dominant rain-producing weather systems and has long been identified as an area vulnerable to climate change-driven rainfall reductions (Archer et al. 2019). From an adaptation perspective, the questions are whether the shift of weather system is permanent and the nature of an appropriate adaptation response.

The city's new postcrisis strategy (Cape Town 2019) acknowledges continued dependence on surface water for 75% of its supplies for at least the next decade, while alternative sources such as reuse, groundwater, and desalination are developed to provide a greater proportion of supplies in the future.

Climate uncertainty is specifically addressed. The program aims to increase the assurance of supply from the system from 98% to 99.5% but acknowledges that there might be "a step change in rainfall due to climate change . . .. If this turns out to be the case, the programme will be both accelerated and expanded."

The strategy also acknowledges that water investments must provide regular supplies during periods of low rainfall: "All water schemes provide insurance against periods of low rainfall, which may become more frequent and more severe as a result of climate change." It recognizes that expensive supplies from reuse and desalination will not be used all the time but concludes: ". . .this will not have been wasteful expenditure. The future is uncertain, and the cost of very severe restrictions is much higher than the cost of insuring against this likelihood."

In this regard, the city's strategy now explicitly addresses climate change. But rather than placing reliance on forecasting future availability, it seeks to increase reliability (through infrastructure investment originally proposed to address variability) to "buy time" for implementing rapid interventions such as desalination if required.

Mombasa, Kenya: Climate-Resilient Designs Constrained by Institutions and Finance

Greater Mombasa, Kenya's second largest city, had a population of approximately three million people in 2018. It is Kenya and the East African region's primary port and is also a center for international beach tourism which complements the country's inland game park complexes.

Water supply is deeply deficient. Water requirements are estimated to be in the region 250 Ml/day, but the city is in a permanent state of water crisis, able to provide less than 30% of this potential amount (Foster et al. 2020). Small supplies from local coastal aquifers are at the limit of their available capacity and face problems of salinization. The city's primary supply comes from two distant groundwater sources, the Mzima Springs (200 km away) and Baricho (100 km).

Unusually, local surface water sources had not been tapped although the need for a new source had already been identified in the early 1990s. The preferred augmentation project, now in preparation, is a dam on the small, local Mwache River.

All five projects in the 2015 World Bank infrastructure resilience study (Cervigni et al. 2015) which included Mwache faced considerable uncertainty about future rainfall under climate change. In none was there even a consensus on whether it would be higher or lower than present-day averages, let alone by how much. This was a particular concern for Mombasa whose rainfall derives from a combination of systems since 61% of the 121 climate future scenarios generated for Mwache showed lower safe yields than the present-day design assumptions.

The World Bank study took a measured response to this systemic uncertainty. It distinguished between project *sensitivity* to climate change and project *vulnerability*; sensitivity refers to the possible reduction in physical "output," while vulnerability considers the impact on economic return. In this view, even if a project's performance is potentially sensitive to climate change, "the project's economic worthiness is not necessarily in question." This would include cases where the project was robust to a high degree of climate variability "and in the bargain, to climate change" so that its benefits and revenues would meet the required criteria.

"It is thus important to distinguish between climate sensitivity and vulnerability" emphasized the report, noting that both sensitivity and vulnerability would depend on the metrics used to assess performance and that factors other than climate change, such as price and demand, could be equally important.

For Mwache, although limited hydrological information was available, the design chosen was considered to have low climate risk since the river flow is considerably in excess of supply requirements and there is a low risk that the dam will not meet its target yield (World Bank 2018b).

Funding has not yet been approved, not due to climate change risks but because the financial and institutional arrangements for the dam's management must still be resolved. Kenya's 2010 Constitution makes water supply the responsibility of local counties. But Mombasa's current supplies come from other counties, and the Mwache Dam itself will be built in a neighboring county.

Current proposals are to establish a regional institution that would manage all the different sources into a single system, providing bulk supplies to the major centers of four counties. This would help to address climate risks since linking different sources into a single system would increase its resilience to climate variability, maintaining supplies even if one source in the system fails. But achieving financial agreement between four counties and national government is proving to be more difficult than finding a climate-resilient dam design.

Windhoek, Namibia: A Climate Secure Source Using Groundwater as Storage

Namibia's capital Windhoek, high in the center of the arid and sparsely populated country, was established in 1892 by German colonists, attracted by the secure source of water from local springs. This groundwater resource served the growing settlement well until the 1960s when, after a series of dry years, it became obvious that additional supplies were needed.

In 1968, a "direct potable reuse" plant was built that treats wastewater for reinjection into the city's main supply. Subsequently expanded, it can now provide 20% of the city's potable supply. An ambitious longer-term plan envisaged a phased regional scheme, the Eastern National Water Carrier (ENWC), drawing from

Southern Africa's third largest river, the Okavango, on Namibia's border with Angola, 700 km from Windhoek. The full project has not yet been completed, but growing demand has been met by intermediate interventions, including the development of new groundwater sources on the ENWC route. Throughout this period, the city's local groundwater complemented other sources.

The use of these multiple sources varies in response to the extremely variable climate, with surface water predominating in wet seasons and greater use of groundwater and reuse in dry periods. Local aquifer recharge has recently been introduced to store surplus water in wet periods, reducing evaporative losses from the surface reservoirs and increasing system yield (Murray et al. 2018).

However, while diversification of sources provides some additional resilience to inter-seasonal climate variability, the assured yield still depends on local natural recharge supplemented by surface water imports. Drought in 2015/2016 showed that despite tightly managed demand, additional supply was needed and attention turned once again to imports through the ENWC. Although capital and operating costs appear to be prohibitive, this would substantially reduce climate risk by diversifying to an unconstrained source.

In this difficult environment, available hydrological tools have consistently provided good estimates of the volumes of water available at different levels of assurance to guide system management. The technical analysis of the recharge and storage potential of the Windhoek aquifers has also provided the basis for implementation. Although the impacts of climate change on regional precipitation and Okavango River flows are uncertain (Hughes et al. 2011), Namibia's requirement is only around 2% of the total flow, suggesting that the objections of environmentalists in Botswana, the downstream riparian, are ill-founded.

The other long-term policy option is to break the "path dependence" since Windhoek was made the country's capital 130 years ago and promote economic development in Namibia's more humid northern region rather than increase the city's water supply. Absent such a radical policy shift, the incremental completion of the ENWC is a rational approach to making Windhoek's water supply more resilient since the scope for further savings through demand-side interventions is limited.

The city's century-long hydrological history and the acute nature of the water challenges that it faces have produced a good understanding of the options available to build systems that are resilient to the risks posed by long periods of multi-season drought. As a result, if adequate financial resources are available, the strategies already adopted to manage current climate variability will enable the city to meet its needs, even under conditions of climate change.

Beira, Mozambique: Sea Level Rise Will Compound Existing Challenges

Like Windhoek, the Mozambican City of Beira faces many climate-related challenges given its location which is regularly hit by tropical storms. Established in 1890 as a port and regional railway hub on low-lying land at the mouth of the Pungwe River, urban water supply is difficult because fresh groundwater is very limited and the river's estuary is saline. While river flow is adequate to meet the city's needs, extensive saline intrusion, exacerbated by growing upstream water use, required its intake 80 km from the city to be moved further 20 km upstream. However, water quality problems continued for a nearby irrigation scheme whose intake the city had shared. While the intake could have been protected by building a weir downstream to obstruct tidal influx or storage dams upstream to maintain minimum dry season flows, these options were rejected on environmental grounds (NORAGRIC 1997).

While climate change impacts are obvious, hydrological analysis has character-ized the salinity dynamics and helped to identify appropriate responses to the impact of increased water utilization and sea level rise. The immediate water management challenges have been well-characterized, and the priority is to coordinate responses to protect economic activity. However, in the longer term, the city's vulnerable location may require radical relocation rather than water management solutions.

Polihali Dam and the Integrated Vaal River System: Sustaining a "Problemshed"

In economic terms, the Integrated Vaal River System (IVRS) is the most important water resource development in Southern Africa. It supplies water to around 20 million people in a region that produces almost 50% of South Africa's GDP and includes the country's administrative capital (Pretoria) as well as Johannesburg, its largest city. If its functioning is impaired by climate change impacts, the entire country will suffer.

The IVRS region lies across the watershed of the country's two largest rivers, the Orange and the Limpopo, which discharge into the Atlantic and Indian Oceans, respectively. By the 1960s, demand for water for domestic and economic purposes had outstripped the reliable flow of the main local source, the Vaal River, a tributary of the Orange River, which is also highly variable. A national policy review (South Africa 1970) recommended a system approach to identify and introduce new sources. This coincided with the emergence of new techniques to undertake the required analysis (Maass et al. 1962).

For four decades, system models guided decisions on expansion and operation (Basson and Van Rooyen 2001) that have successfully sustained water security in the expanding system. Throughout this period, the objective was to manage the extreme climate variability, with drought risk as the key metric since there was little useful information about potential climate change risks.

An important element of the system is the Lesotho Highlands Water Project (LHWP), a multiphase scheme which diverts water that flows south from Lesotho in the Orange River, northward to its Vaal River tributary. In terms of a binational Treaty, South Africa uses Lesotho territory to discharge the waters closer to the centers of demand, with considerable cost-savings. Climate variability and change are addressed in the Treaty only insofar as provision for joint action in the event of *force majeure*, covering "disturbance due to an extreme hydrological or other natural

event, including extreme drought, and affecting the delivery of water to South
Africa." Similarly, while the 1998 Appraisal Report on Phase 1B of the project
(which was successfully completed in 2002) makes no explicit mention of climate
variability and change, it locates the purpose of the project as "reducing drought
risk" in the South African system.

The LHWP's further phases are seen as low-risk, least-cost alternatives to augment the system's capacity. Phase 2, the construction of the Polihali Dam on another
Orange River tributary, was included in the World Bank's "RDM study" (Cervigni
2015). As with the Mwache Project in Kenya (see above), the study considered
whether alternative configurations might be less economically vulnerable if climate
change reduced project performance.

Noting considerable climate uncertainty, the study suggested an alternative
design could reduce the risk of financial losses due to climate change by up to
30%. This only considered assumptions about water prices and demand and not the
wider societal costs of supply failure, a conservative approach since, for many
societies, higher water costs are preferable to the costs of unexpected supply failures
(as is formally stated in the new Cape Town water strategy (Cape Town 2019)).

Subsequent studies concluded that the Polihali Dam would meet its intended
delivery targets under a wide range of climate scenarios with deficits occurring only
in the very driest 16 of 122 scenarios. The IVRS also illustrates the benefits of a
"problemshed"-based analysis (Mollinga 2020) which encompasses the actual physical and institutional boundaries within which water is managed not simply an
individual "watershed" approach. Whereas hydrological models are based on the
natural "watersheds" before human activity intervenes, water management requires
models that reflect the interaction between human activities and the natural system.
The IVRS shows how hydrological tools can effectively be applied to manage
substantial climate variability in complex multi-basin systems.

Zambezi River Coordination: Drought Risks, Discount Rates, and Batoka Dam Hydropower

The final case, the proposed Batoka Dam in the transnational Zambezi River Basin,
illustrates the challenges of evaluating climate change's impacts in a complex system
and uncertain economic and institutional context. In a basin that already suffers
extreme climate events, the implications of climate change for hydropower expansion are a concern. Hydropower is the main source of electricity for the riparian
countries and offers further opportunities to support their socioeconomic development. The Batoka Dam project illustrates the challenges of identifying adaptive
approaches that optimize the benefits that hydropower can provide while minimizing
the risk.

The history of Zambezi hydropower is as much about national and regional
politics as hydrology or economics. Its potential was recognized early in the
twentieth century, but it was only in the 1950s that it was decided to construct the
Kariba Dam, creating what is still the world's largest dam reservoir by volume. Built

by the Government of the Federation of Rhodesia and Nyasaland, Kariba's initial installed capacity of 1300 MW was sufficient to meet the needs of the Federation. It was controversial because a smaller project could have been built on the Zambezi's Kafue tributary in Northern Rhodesia, closer to the mines that were the main centers of electricity demand. But political imperatives supported the larger project to reinforce the Federation although Northern Rhodesia, later Zambia, withdrew in 1963.

Mozambique's Cahora Bassa Dam, with an installed capacity of 2075 MW, had similar political drivers to Kariba; it used cheap electricity to reinforce the political relationship between the Portuguese colonial power and South Africa. This relationship was even shorter-lived – Mozambique became independent in 1975, just as Cahora Bassa was completed and the reservoir began to fill.

With the development in Zambia of the 900 MW Kafue Gorge Upper Dam and some smaller projects, there is now almost 5,000 MW of installed hydropower generation capacity in the basin with an average total energy production of around 30,000 GWh/year, of which 23,000 is "firm energy," available at high reliability and very low cost, since most initial construction loans have long been repaid. This economic benefit of large water infrastructure is seldom accounted for although the resulting "path dependency" is often commented upon (Haasnoot et al. 2019).

The total energy production could be doubled to around 60,000 GWh/year through the extension of existing facilities and the construction of new dams. This would meet most of the current electricity demand of the eight riparian states although optimal production will require hydrologically informed cooperation between operators. The resilience of new schemes to climate variability and change has come under particular scrutiny as international development finance institutions consider their financial viability.

While individual projects have been assessed in some detail, it is important to understand how they would function in a future system, which includes irrigation and urban uses, under conditions of climate change. At present, evaporation from the Kariba and Cahora Bassa lakes is by far the largest "consumptive use" of Zambezi water, accounting for 85% of the 12.5 km^3 average "consumption"; all other uses (agriculture, urban, and industrial) account for just 1.9 km3, just 2% of the available runoff.

With the growth of population and economies, water consumption will reduce flows, impacting on hydropower production. A 2010 "Multi-Sector Investment Opportunity Analysis" (MSIOA) investigated the potential impacts of such new developments and the benefits for the riparian countries of coordinating their development plans, to reduce uncertainty about hydropower potential (World Bank 2010). The study included an economic assessment tool and considered possible climate change impacts. Its main finding was that coordinating investments and infrastructure operation could significantly increase the economic benefits compared to stand-alone project development. Beyond the hydrological uncertainties, the political challenge is to achieve an equitable sharing of the costs and benefits.

However, economic analysis is particularly challenging for investments such as water resource infrastructure which have a high initial cost but yield benefits over a very long period. The imputed value of long-term production depends heavily on assumptions about future prices as well as the "discount rate" applied in the analysis, and there is limited consensus about the appropriate approach, not least because the choice is usually determined by policy objectives (Fankhauser and Stern 2016) in which climate change adaptation strategies are not always adequately reflected.

These issues were illustrated in the analysis of Batoka in the 2015 RDM study (Cervigni 2015). The Batoka site is on the main stem of the Zambezi, downstream from Victoria Falls and upstream from Kariba. It has limited storage and will depend on natural river flows which, although moderated somewhat in the extensive wetlands of the upper catchment, are very variable and subject to drought, making it particularly vulnerable to climate change (Harrison and Whittington 2002).

In this context, the RDM study noted that:

"In drier futures, smaller facilities yield higher net benefits, as less investment is underutilized during the dry periods. In wetter futures, larger facilities that can better take advantage of high flow periods yield higher net benefits."

Batoka it found,

"shows significant sensitivity to climate change with up to a 33 percent decrease or a 15 percent increase in average power production." (Cervigni 2015)

Over 30 years, the difference between the worst- and best-case scenarios was estimated at around US$4 billion, in 2015 values, based on average energy prices in the Southern African Power Pool. But in a dry period, electricity prices would rise as all supplies would be reduced in a system that was dependent on hydropower. While in 2015 this effect was not yet evident because the region still had coal-fired alternatives, over the next five decades, coal-fired power will be constrained, and carbon taxes will increase its price, further supporting hydropower prices.

Even without these considerations, Batoka showed that "economics of projects of this type could be highly sensitive to the price of power" (Cervigni et al. 2015). Once again, it was emphasized that a project that was financially sensitive to climate change impacts was not necessarily financially vulnerable.

These economic studies are highly dependent on the discount rates chosen. The RDM study used a policy discount rate of 3% for assessing social policy objectives. But for Batoka, both the MSIOA and the RDM analyses used a "finance discount" rate of 10% (World Bank 2010) to ensure adequate financial returns to repay loans. The danger is that using present-day markets to estimate long-run prices ignores the potential increase in the value of output as a result of climate change policies, devalues the long-term benefits of cooperation, and thus does not adequately reflect the policy goals of climate change mitigation.

A final gap in these attempts to integrate policies for climate resilience more effectively fully into hydrological and economic analysis has been the failure to consider the "water use efficiency" of energy generation in the Zambezi Basin. Presently, over 10% of the Zambezi's flow is lost to evaporation from the two large hydropower reservoirs. Full development of the basin will not require the significant expansion of storage to manage hydrological variability, and the proposed cascade of three dams in the lower Zambezi will benefit from the hydrological security provided by the Kariba and Cahora Bassa reservoirs upstream. While the MSIOA notes with concern that the output of those projects could vary between +9% and − 13% by 2100 (World Bank 2010), it makes no allowance for value inherent in increasing "water use" efficiency by increasing the energy produced per unit of water.

These analyses of the Batoka Dam and the broader strategies to ensure optimum management of the Zambezi Basin's waters show that the challenges of assessing the economic benefits and risks of large water investments are at least as complex as those of managing the hydrological uncertainties.

Discussion: Principles to Guide Adaptation under Uncertainty

The cases presented illustrate the contexts within which major water resource investment decisions are taken and the limited contribution that can be made by emerging climate change science beyond the charting of the range of future uncertainty. Nevertheless, the focus on potential climate change impacts does provide useful guidance for the planners, developers, and operators of major water resource infrastructure working in "difficult hydrologies."

Principles

From these cases, some general principles can be identified that could help water resource managers to develop coherent adaptive strategies for the projects or systems for which they are responsible and assist them in communicating their approaches to their wider communities. In addition, some more operational priorities emerge that merit mention.

Uncertainty

Perhaps the most important principle to emerge from the cases is that uncertainty remains the dominant feature of potential climate change impact on water resource projects. This uncertainty must be recognized and reflected in analytical approaches, strategy development, and project design. Uncertainty does not mean that stochastic methods cannot be used to detect − and project − trends. For instance, where projections of precipitation are made using multiple GCMs and the range of future uncertainties can be constrained, stochastic methods are already usefully applied. More serious challenges arise where relevant variables are not well understood or constrained as is the case with modeling future rainfall-runoff in conditions where both land cover and precipitation intensities are both poorly understood.

Precautionary

The precautionary principle is usually understood to constrain developments until their impacts are understood. In the climate change context, a precautionary approach will ensure that decisions are taken in time to address possible future contingencies. This is explicitly stated in Cape Town's post-disaster water strategy which commits to make new investments which will not be used unless there is a dry period in which they are needed. This reflects the high costs imposed on the wider society by supply failure. Projects with lengthy implementation periods such as the IVRS augmentations should be implemented on a precautionary basis to avoid inefficient crisis responses that characterized many water sector investments. A precautionary approach would also introduce supply restrictions earlier rather than later although political resistance is always a factor in this – recent extreme power restrictions in Zimbabwe resulted, in part, from earlier decisions to continue generating at full capacity at Kariba, contrary to drought operating rules.

Path Dependence

A concern frequently raised about water resource developments of all kinds is that they commit their societies to particular development paths and close off other options. The difficulty with these discussions is that they frequently involve larger political issues about which there is little consensus. In large measure, Windhoek's current challenges are the result of century-old colonial spatial planning decisions; the option of refocusing economic development in a more propitious location is an important alternative. Similarly, South Africa's IVRS reflects the fact that the mining economy promoted extensive development in a location with limited water resources and that coastal areas would be easier to serve. Beira's location, which also served colonial objectives, is very vulnerable to current climate, and vulnerabilities will be aggravated by climate change. But these are not issues on which water managers can determine policy. At best, they can ensure that the concerns and constraints are well understood and thus influence wider development policy debates.

Flexibility

In response to uncertainty about future trends, it is helpful if strategies to meet future needs are sufficiently flexible to accommodate different futures. This principle has already been explicitly adopted in a number of the case study projects: Cape Town has identified a range of options that could rapidly be implemented in the event of a "step change" in climate; the ENCW scheme serving Windhoek can be implemented in a sequence of steps. Similarly, each step in the past and future development of South Africa's IVRS involved a choice from a set of different options.

System-Wide Approaches and Problemsheds

The importance of conceptualizing and operating linked infrastructure as a system rather than as individual units brings many benefits. The values of these benefits are explicitly determined in the Zambezi Basin MSIOA but are also evident in the IVRS, Windhoek, and, at a nascent stage, Mombasa and Cape Town's new strategy.

Linkages

The linkage of different sources and demand centers provides a variety of resilience benefits that can greatly assist adaptation to changing climate conditions. A benefit that could be achieved by operating Zambezi hydropower as a coordinated system is that extreme events are often not correlated. While in the south Victoria Falls and Kariba were running dry, Cahora Bassa which draws more resources from the north had to open its flood gates to manage inflows. South Africa's IVRS has already demonstrated that multiple sources provide greater assured yields than the sum of their "stand-alone" yields and Windhoek's situation is similar. Proposals for Kenya's Mwache Dam to form part of an integrated bulk system would replicate this approach.

Diversification

Diversification of sources provides further important resilience benefits and should be a guiding principle for climate change adaptation. Windhoek's surface water/groundwater mix has enabled it to withstand repeated drought crises. Cape Town's recent experience highlighted the risk of relying solely on surface water supplies, which although linked all derived their inflow from a single small catchment area. Alternative sources had not been developed because, individually, their unit costs were higher, but this calculation omitted consideration of the value of additional reliability that could be gained from diversification. Care must be taken, however, to ensure that apparently diversified sources are not correlated. Thus, while reuse is a useful option, it depends on the availability of wastewater which, in crisis situations, may be constrained.

Storage Vs Yield

It is widely recognized that storage is an integral element of any substantial water resource system, crucial to ensure adequate flows in dry seasons. However, it is often considered solely as part of river management infrastructure. Both Windhoek and the IVRS use strategic storage fed slowly by wet season surpluses from external sources to build a reserve that can be drawn down to provide a guaranteed supply during dry periods. This contributes significantly to their reliance against drought and other climate impacts at lower cost than would be required for the development of a new source.

Valuing Resilience and Reliability

One of the features of both water and power projects in Southern and Eastern Africa is how little recognition is given to the financial and economic benefits of reliability and resilience of supply. This was forcefully illustrated by the economic losses suffered by Cape Town due to 3 years of water restrictions. But the economies of countries dependent on Zambezi hydropower have also been significantly constrained by the power cuts and rationing, and Windhoek's economic activity has been curtailed by water shortages. There is a need for more nuanced analysis of the funding of schemes that enhance resilience and reliability of supply since these values are often not included in user charges and benefit a wider public than the direct users.

Operational Priorities

Institutional Capacity Will be Critical

Given the intensification of hydrological analysis and decision-making, water management institutions will have to ensure that they have the capabilities to respond to the emerging challenges. This will require appropriate structures as well as the development and retention of the specialist skills, often in collaboration with specialized institutions of higher education.

Adequate Finance Is Required

In all the water supply cases, current investment has been insufficient to achieve resilience to current climate variability. While this is often viewed as a problem that can be resolved by establishing effective markets for water (and also to power), the benefits of assured supplies spill far beyond the direct users to the wider society although they may only be fully realized in the longer term. This means that the full value of the water project outputs cannot be captured by prices alone and different financial arrangements are required that reflect the long-run economic benefits of secure supplies. Where water projects, notably hydropower, directly mitigate climate change by reducing greenhouse gas emissions, this should be reflected in financial terms. Current "green finance" proposals will only be of assistance if loan conditions address these structural constraints.

Communication to Achieve Sufficient Consensus

Water resource development occurs to a greater or lesser extent in the public domain and is thus subject to public scrutiny and political decision. One of the challenges faced by sector managers is that they address complex issues about which public perceptions may differ widely. Water management institutions need to develop and sustain communication with their stakeholder communities to ensure their support for demand-side interventions as well as for investments on the supply side.

Information for "Dynamic Stationarity"

The dynamic nature of hydrological systems under climate change means that analysis will no longer rely on standard sets of reference data from a fixed time period. The values of many hydrological variables will be changing in unpredictable ways that will require hydrological data to be updated on a regular basis. This will require an intensification of data collection and management. As one group of commentators remarked, "in a nonstationary world, continuity of observations is critical" (Milly et al. 2008).

Conclusions

The principles that emerged from this review support the initial hypothesis that adaptation to climate change in the water and hydropower sector will best be achieved by following strategies that address current climate variability. All of the principles enunciated are equally appropriate to current climate variability or climate change.

The proviso introduced by including climate change is that such interventions should recognize that uncertainties under future climate change are likely to be greater than at present. This reinforces the need for options that allow greater flexibility of approach, but it also highlights the need for a precautionary approach that will see interventions made earlier rather than later, as Cape Town found to its cost.

The idea that interventions should be made sooner rather than later will often meet opposition because it will be seen to reinforce current development trajectories and path dependence. The reason that alternative trajectories are not followed is usually that they are constrained by politics and economics. Even if the intent is to allow economic and social disruption to drive new trajectories, creating water and power crises is unlikely to create the conditions for change. Given the challenges posed by climate change, deliberate disruption of societies and economies is a high risk strategy with little certainty about the likely outcomes.

One consistent finding is that the challenges of planning, implementing, and operating large water resource schemes under difficult hydrologies will become more complex as climate change impacts intensify. While traditional hydrological tools can continue to be used, "dynamic stationarity" will require more intensive data collection, more frequent reviews and runs of models, and greater effort to communicate the findings and their implications to the wider community whose consent, trust, and support will be required for decisions. In most countries, this will require more capable organizations, better funded, staffed, and equipped than at present.

In all of this, the primary constraint to adapting to the impacts of climate change by building more resilient and less vulnerable systems will continue to be financial. A particular challenge for the water sector's practitioners will thus be to persuade their communities that additional support for planning and managing their countries' water will insure them against the larger costs of water and power supply failures.

References

Adler RF, Gu G, Sapiano M, Wang JJ, Huffman GJ (2017) Global precipitation: means, variations and trends during the satellite era (1979–2014). Surv Geophys 38(4):679–699

Archer E, Landman W, Malherbe J, Tadross M, Pretorius S (2019) South Africa's winter rainfall region drought: a region in transition? Clim Risk Manag 25:100188

Bailey AK, Pitman WV (2016) Water resources of South Africa, 2012 Study (WR2012), WR2012 study executive summary. WRC Report TT, 683. Water Research Commission, Gezina, p 16

Basson MS, Van Rooyen JA (2001) Practical application of probabilistic approaches to the management of water resource systems. J Hydrol 241(1–2):53–61

Blochl et al (2019) Twenty-three unsolved problems in hydrology (UPH) – a community perspective. Hydrol Sci J 64(10):1141–1158. https://doi.org/10.1080/02626667.2019.1620507

Brekke LD, Kiang JE, Olsen JR, Pulwarty RS, Raff DA, Turnipseed DP, Webb RS, White KD (2009) Climate change and water resources management – a federal perspective: U.S. Geological Survey Circular 1331, 65 p. Also available online at http://pubs.usgs.gov/circ/1331/

Briscoe J (2010) Practice and teaching of American water management in a changing world. J Water Resour Plan Manag

Cape Town (2019) Our shared water future: Cape Town's water strategy, City of Cape Town

Cervigni R, Liden R, Neumann JE, Strzepek KM (2015) Enhancing the climate resilience of Africa's infrastructure: the power and water sectors. Overview booklet. World Bank, Washington, DC

Chahine M (1992) The hydrological cycle and its influence on climate. Nature 359:373–380. https://doi.org/10.1038/359373a0

Chan JC (2006) Comment on "Changes in tropical cyclone number, duration, and intensity in a warming environment". Science 311(5768):1713–1713

Conway D (2005) From headwater tributaries to international river: observing and adapting to climate variability and change in the Nile basin. Glob Environ Chang 15(2):99–114

Fankhauser S, Stern N (2016) Climate change, development, poverty and economics (presented at) The State of Economics, the State of the World conference. World Bank, Washington, DC

Foster S, Eichholz M, Nlend B, Gathu J (2020) Securing the critical role of groundwater for the resilient water-supply of urban Africa. Water Policy 22(1):121–132

Gangrade S, Kao SC, Naz BS, Rastogi D, Ashfaq M, Singh N, Preston BL (2018) Sensitivity of probable maximum flood in a changing environment. Water Resour Res 54(6):3913–3936

Giorgi F, Raffaele F, Coppola E (2019) The response of precipitation characteristics to global warming from climate projections. Earth Syst Dynam 10:73–89. https://doi.org/10.5194/esd-10-73-2019

Greve et al (2019) The aridity index under global warming. Environ Res Lett 14:124006

Grey D, Sadoff CW (2007) Sink or swim? Water security for growth and development. Water Policy 9(6):545–571

Haasnoot M, van Aalst M, Rozenberg J, Dominique K, Matthews J, Bouwer LM, Kind J, Poff NL (2019) Investments under non-stationarity: economic evaluation of adaptation pathways. Clim Change:1–13

Harrison GP, Whittington HBW (2002) Susceptibility of the Batoka gorge hydroelectric scheme to climate change. J Hydrol 264(1–4):230–241

Hegerl GC, Brönnimann S, Cowan T, Friedman AR, Hawkins E, Iles C, Müller W, Schurer A, Undorf S (2019) Causes of climate change over the historical record. Environ Res Lett 14(12):123006

Herold N, Behrangi A, Alexander LV (2017) Large uncertainties in observed daily precipitation extremes over land. J Geophys Res-Atmos 122:668–681

Hughes DA, Kingston DG, Todd MC (2011) Uncertainty in water resources availability in the Okavango River basin as a result of climate change. Hydrol Earth Syst Sci 15(3)

Hüttl RF, Bens O, Bismuth C, Hoechstetter S (2016) Society-Water-Technology: a critical appraisal of major water engineering projects. Springer Nature, p 295

Knutson T, Camargo SJ, Chan JC, Emanuel K, Ho CH, Kossin J, Mohapatra M, Satoh M, Sugi M, Walsh K, Wu L (2020) Tropical cyclones and climate change assessment: Part II: projected response to anthropogenic warming. Bull Am Meteorol Soc 101(3):E303–E322

Kraemer BM, Seimon A, Adrian R, McIntyre PB (2020) Worldwide lake level trends and responses to background climate variation. Hydrol Earth Syst Sci 24(5):2593–2608

Loucks DP, Van Beek E (2017) Water resource systems planning and management: an introduction to methods, models, and applications. Springer

Maass A, Hufschmidt MA, Dorfman R, Thomas HA Jr, Marglin SA, Fair GM (1962) Design of water-resource systems: new techniques for relating economic objectives, engineering analysis, and governmental planning. Harvard University Press, Cambridge, Mass

Malherbe J, Engelbrecht FA, Landman WA, Engelbrecht CJ (2012) Tropical systems from the Southwest Indian Ocean making landfall over the Limpopo River basin, southern Africa: a historical perspective. Int J Climatol 32(7):1018–1032

Matalas NC (1997) Stochastic hydrology in the context of climate change. Clim Chang 37(1):89–101

Milly PCD, Betancourt J, Falkenmark M, Hirsch RM, Kundzewicz ZW, Lettenmaier DP, Stouffer RJ (2008) Stationarity is dead: whither water management? Earth 4:20

Mollinga PP (2020) Knowledge, context and problemsheds: a critical realist method for interdisciplinary water studies. Water Int. https://doi.org/10.1080/02508060.2020.1787617

Moss ME, Tasker GD (1987) The role of stochastic hydrology in dealing with climatic variability. In: The influence of climate change and climatic variability on the hydrologic regime and water resources (Proceedings of the Vancouver symposium, August 1987), vol 168. IAHSPubl, pp 201–207. Accessed at http://hydrologie.org/redbooks/a168/iahs_168_0201.pdf

Muller M (2018) Lessons from Cape Town's drought. Nature 559(7713):174–176

Muller M, Biswas A, Martin-Hurtado R, Tortajada C (2015) Built infrastructure is essential. Science 349(6248):585–586

Murray R, Louw D, van der Merwe B, Peters I (2018) Windhoek, Namibia: from conceptualising to operating and expanding a MAR scheme in a fractured quartzite aquifer for the city's water security. Sustaina Water Resour Manage 4(2):217–223

NORAGRIC (1997) Environmental assessment of the Mozambique National Water Development Project. World Bank. Accessed at http://documents1.worldbank.org/curated/en/165791468779949188/pdf/multi-page.pdf

Pillay MT, Fitchett JM (2019) Tropical cyclone landfalls south of the tropic of Capricorn, southwest Indian Ocean. Clim Res 79(1):23–37

Sidibe M, Dieppois B, Mahé G, Paturel JE, Amoussou E, Anifowose B, Lawler D (2018) Trend and variability in a new, reconstructed streamflow dataset for west and Central Africa, and climatic interactions, 1950–2005. J Hydrol 561:478–493

South Africa (1970) Commission of enquiry into water matters, Report of the commission of enquiry Into water matters, vol 34. Government Printer, Pretoria

Strzepek K, McCluskey A, Boehlert B, Jacobsen M, Fant C IV (2011) Climate variability and change: a basin scale indicator approach to understanding the risk of climate variability and change: to water resources development and management. World Bank, Washington, DC, pp 1–139

Trelease FJ (1977) Climatic change and water law. In: National research council, climate, climatic change, and water supply. National Academies, Washington, DC

Trenberth KE (2011) Changes in precipitation with climate change. Clim Res 47(1–2):123–138

Ukkola AM, De Kauwe MG, Roderick ML, Abramowitz G, Pitman AJ (2020) Robust future changes in meteorological drought in CMIP6 projections despite uncertainty in precipitation. Geophys Res Lett e2020GL087820

Vaze J, Post DA, Chiew FHS, Perraud JM, Viney NR, Teng J (2010) Climate non-stationarity-validity of calibrated rainfall–runoff models for use in climate change studies. J Hydrol 394(3–4):447–457

Venter ZS, Cramer MD, Hawkins HJ (2018) Drivers of woody plant encroachment over Africa. Nat Commun 9(1):1–7

Vogel RM (2017) Stochastic watershed models for hydrologic risk management. Water Secur 1:28–35

Woodhouse P, Muller M (2017) Water governance – an historical perspective on current debates. World Dev 92:225–241

World Bank (2010) The Zambezi River basin a multi-sector investment opportunities analysis (Volume 1) Summary report. World Bank, Washington, DC

World Bank (2018a) Assessment of the state of hydrological services in developing countries. World Bank, Washington, DC

World Bank (2018b) Economywide and distributional impacts of water resources development in the coast region of Kenya: implications for water policy and operations. World Bank, Washington, DC

Zareian MJ, Eslamian S, Gohari A, Adamowski JF (2017) The effect of climate change on watershed water balance. In: Mathematical advances towards sustainable environmental systems. Springer, Cham, pp 215–238

7

Farmers' Adoption of Climate Smart Practices for Increased Productivity in Nigeria

B. E. Fawole and S. A. Aderinoye-Abdulwahab

Contents

Abstract

In a bid to reinforce the efforts of agricultural professionals within the domain of climate change studies and with particular emphasis on rural farmers in Nigeria, this chapter explores the mechanics for adoption of climate smart agricultural practices among rural farmers for an increased agricultural productivity. Climate-Smart Agriculture (CSA) is paramount to the success of farming activities today in the face of the menace of the impact of climate change. Climate Smart Agricultural

B. E. Fawole
Department of Agricultural Extension and Rural Development,
Federal University, Dutsinma, Nigeria
e-mail: bfawole@fudutsinma.edu.ng

S. A. Aderinoye-Abdulwahab (✉)
Department of Agricultural Extension and Rural Development,
Faculty of Agriculture, University of Ilorin, Ilorin, Nigeria
e-mail: aderinoye.as@unilorin.edu.ng

Practice (CSAP) is one of the major keys that agricultural development approaches aimed at; to sustainably increase productivity and resilience, while also reducing the effects; as well as removing emissions of greenhouse gases. It is pertinent to note that most of the CSAPs adopted by the rural farmers in this study are conservation agriculture, use of organic manure, crop diversification, use of wetland (Fadama), planting of drought tolerant crops, relocation from climate risk zones, prayers for God's intervention, and improvement on farmers' management skills. This study divulged and showcased the import of CSAP in boosting agricultural yield and also highlights the bottlenecks inhibiting agricultural farming practices such as lack of practical understanding of the approach, inadequate data and information, lack of suitable tools at local and national levels, supportive and enabling policy frameworks, and socioeconomic constraints at the farm level. The study concluded by recommending an aggressive awareness and mobilization campaign to boost the adoption of CSAPs in Nigeria.

Keywords

Rural areas · Agricultural risks and productivity · Policy · Relocation · Greenhouse gases

Introduction

Agriculture plays a huge role in Africa's economy with about 70% of the continent's population practicing subsistence farming for their livelihood while they equally live in the rural areas. Hence, the significance of this sector in providing employment and motivating economic growth in a developing nation such as Nigeria cannot be undermined. Agriculture mainly hinges on environmental factors and any prolonged variations in average weather conditions can have significant impacts on production. Additionally; agriculture as practiced in the continent, is highly characterized by land tenure insecurity which continues to hamper agricultural growth for millions of smallholder farmers; while poor soil fertility, degraded ecosystems and climate variability are among other problems associated with agriculture in Africa (USAID 2016). Climate change adaptation has to do with any spontaneous and/or premeditated action/mechanism adopted to deal/cope with the impacts of, or reduce vulnerability to, a changing climate.

Adaptation should address immediate problems and anticipate future changes in order to mitigate adverse effects. The strategies applied in adaptation can be preventive or reactive while all the methods taken should aim at enhancing resilience and reducing vulnerability (IPCC 2012). There is a dire need to deliberately advance a rigorous implementation of the identified strategies for adaptation and mitigation. This is in a bid to minimize the effects of climate change in order that development in agriculture can be encouraged. Mitigation refers to measures that may either reduce the increase in greenhouse gases (GHGs) emissions or increase terrestrial storage of carbon, while adaptation refers to all the responses to climate change that may be used to reduce vulnerability (Ifeanyi-obi and Nnadi 2014). Indeed, there have been deliberate efforts by Heads of States to integrate approaches that have been employed to solve the impacts of climate variability. Evidence is seen in the authorization of governments to support at least twenty-five million farming households to adopt Climate Smart Agriculture (CSA) by 2025. (African Climate Smart Agriculture Summit 2014).

The concept of CSA, which came into limelight in 2010, is defined as a form of agriculture that increases agricultural production and income; enhances adaptation and build resilience to climate change impact. CSA also helps to reduce the emissions due to Greenhouse gases (GHGs) and enhances the attainment of global food safety (World Bank 2010; FAO 2014). Climate-smart ideas include activities that would ensure minimum tillage; application of organic agricultural practices rather than inorganic methods that would further deplete the soil in the long term; crop rotation and mulching to reduce evapotranspiration; and composting and planting of legumes and cover crops which assist in moderating the long-term effects of climate change by giving an enduring environmental strength to the soil (Olorunfemi et al. 2019; Ojoko et al. 2017). This chapter therefore sets to explore the CSAPs available, the benefits accruable from practicing CSA in order to increase productivity and the challenges hindering their adoption in Nigeria.

Adoption Theory as a Conceptual Framework

The choice of adopting a technology is preceded by a number of certain mental processes which include: Awareness, Interest, Evaluation, Trial, and Adoption. According to Ekong (2010), adoption process is the mental procedure from the first instance of learning of an idea/technology to the final stage of taking a decision to use it. The farmer's choice to accept or reject a technology is based on their perception of those stages earlier mentioned. This implies that a farmer can only decide to use a technology only if he was aware of it; develops interest in it; and possibly, after he must have tried it and becomes convinced that the technology is beneficial to him.

Awareness stage: At this first stage, the farmer is barely conversant with the existence of an innovation; it could be that he only became cognizant of the technology and may be deficient in the technicalities such as how to use the innovation, the cost; as well as the benefits of the innovation. The farmer may learn of the existence of the innovation through friends, relatives, mass media, extension agents or local cooperative organization. After being aware of the innovation, the farmer might crave for more information. He may want to understand how the invention works and also the benefits and usefulness. Consequently, the farmer can take the next step of making deep inquiries in order to access adequate information that would help him in reaching a decision.

Interest stage: during this second stage the farmer asks for detailed information concerning the technology. It is only normal that the farmer relies partly on their source (through which they became aware) of information to deepen their knowledge about it. The farmer can also search for better understanding from other sources which hopefully spurs them to become interested in the technology. This makes the farmer to develop interest by asking further questions and collecting information about the new idea or new product to assess its potentialities. These line of questioning can make the farmer to become curious and begin to wonder how much he can benefit should he decide to try the idea/technology.

Evaluation stage: this is the stage where the farmer mentally appraises the applicability of the technology in light of the solution to his own circumstances. The farmer makes a mental assessment by asking several questions such as: Is it/this possible? Who do I know that has done it? What was their outcome? Can I also do it? What happened the last time someone tried it? The responses to these questions and many more that the farmer might have mentally asked would be the determinant of whether he would go ahead to try it or not? If the evaluation produces a negative result, the farmer could stop showing interest in the innovation and may eventually reject it. However, if the responses he got from his mental assessment were favorable, he may then be set to try the innovation practically.

Trial stage: this is the stage where the farmer essentially implements or applies the innovation on a small scale under his personal circumstances and managerial competence. The farmer could entirely handle the trial stage all alone or they may seek the help of change agents and sales promoters. In this stage, the farmer raises the question: how am I supposed to do it? The farmer uses the innovation on a trial/experimental basis on a portion or segment of their farm. The result, positive or negative, would then determine whether the farmer would take or discard the technology.

Adoption stage: The final stage of adoption is when the farmer eventually approves the innovation and has decided to continue using it on a full scale. In this stage, the farmer extends the use of the technology to their whole plots of land. Adoption involves an enduring acceptance and repeated utilization of the technology with the hope that it will help to ease a difficult farm situation. This suggests that a farmer may stop or discontinue the use of such innovation if subsequent outcomes do not prove reliable.

According to Adekoya and Tologbonse (2011), a prospective adopter has a number of characteristics to consider before concluding on whether to adopt an innovation or not. These characteristics are as listed:

1. **Relative advantage**: A technology must be perceived by the farmer as being better than the one currently in use. This can be measured in terms of profitability, low cost, and time saving ability. It could also be in terms of factors relating to social-prestige, convenience and satisfaction. The point that matters in this case is the discernment of advantage as perceived by the adopter.
2. **Compatibility**: This means that the new idea being professed is in consonance with the obtainable values, beliefs, ancient times experiences, and desires of prospective adopters. Hence, any innovation that does not conform with the standards of a community may not be accepted as fast as one that is compatible.

3. **Complexity**: Some technologies are easily and willingly understood by majority of farmers in a community; whereas, others are more difficult and will take more time before they get accepted. Therefore, new ideas should be comparatively easy to comprehend and utilize when compared to existing ones as innovations that are simple to comprehend are accepted more rapidly.

4. **Trial-ability**: An innovation should normally be easily adaptable. What this implies is that, it will be possible to try a technology on a small scale. Any innovation that can be experimented on an installment arrangement will usually be more readily accepted than ideas that are not separable.

5. **Observability**: Farmers would normally want to be able to see the outcome of a new idea so that they could be able to discuss such results with fellow farmers. If outcomes of innovation are readily seen by farmers, then the chances of accepting such innovation become higher; whereas if there are no clear observable outcomes, it may be difficult for such ideas to be replicated and accepted.

6. **Divisibility**: It is a great idea for a technology to be divisible; this means that the innovation can be tested with small units before buying the whole gadget. Purchase of tractors or harvesters is a one-time major investment and part of it cannot be bought in order to try it out in phases. However, such technology can be hired for initial use before deciding on whether to purchase. A farmer can however, purchase and utilize a little quantity of improved seeds, fertilizer or herbicides.

7. **Accessibility**: Innovations that are easily available are more readily accepted. In this case, it is not advisable for a change agent to push for the distribution of farming inputs that farmers either cannot afford or for which infrastructure does not exist; as this would be an indication that the technology is not accessible.

8. **Visibility**: The results of the mode of operation of a given innovation should be easily seen, i.e., visible. A farmer can see a new spraying pump which has been made easier to handle; hence the likelihood of adoption becomes greater.

9. **Cost**: It has been consistently observed that innovation may not be easily adopted if it is perceived to be costlier than what currently exists; it thus goes to say that a new idea may have a relative advantage over an old one; but could meet up with resistance if the currently used practice is a lot cheaper. Generally, the higher the cost of an innovation, the more slowly it is adopted.

However, Rogers (2003) as cited by Ray (2011) argued that the choice to accept any part of an innovation depends on a procedure which is known as the innovation-decision process. Innovation-decision is conceptualized as having five stages and each of these stages can only take place after a period of time has passed. The stages are:

1. Knowledge stage: One is believed to have acquired knowledge the very moment a person is exposed to the technology reality and has achieved some perception of how such knowledge functions. The meaning of this is that knowledge role is primarily cognitive or knowing. People who acquire knowledge at an early stage; are regarded as generally more highly educated and are perceived as having higher social status; they are also more open to both mass media and interpersonal channels of communication, and they have more contact with change agents.

2. Persuasion stage: One would have passed through the persuasion stage once a person forms an opinion about the new practice/idea; though this could only happen after going through several persuasive messages from development workers, sales promoters and close associates. The persuasion function is mainly connected or related to feelings and is thus; a stage laden with so much emotional attachment and/or factors. Hence, it is expected that the potential adopter may turn out to be more psychologically involved with the innovation and would actively seek information about the innovation/new practice. It is at this stage that the would-be adopter forms a good or bad attitude/impression towards the technology.

3. Decision stage: After being persuaded about the usefulness or benefits of an innovation, one would normally engage in activities which may culminate into either of two options; that is whether to take or discard the technology. Having observed that other people use the innovation, the potential adopter can choose to adopt or refuse the technology. People would adopt an innovation only if they are convinced of its usefulness and compatibility with their values and income.

4. Implementation stage: Implementation happen as soon as a person or other decision-making unit place a technology into use. At this stage the individual is generally concerned with where to get the innovation, how to use it and what operational problems will be faced and how those problems could be solved. Implementation might entail alteration in management of the venture and/or adjustment in the technology, to go well with the exact needs of the individual who accepts it.

5. Confirmation stage: This is the last stage in the innovation process when an individual finally accepts or rejects a new practice. Confirmation culminates in repeated use. Individuals seek reinforcement for the adoption decision they have made. The consequence is that they may exhibit continued adoption or exhibit discontinuance. Meanwhile, it is noteworthy to mention that the entire innovation-decision process has a sequence of options at every stage. For instance, in the knowledge function, the would-be adopter would have to decide on which of the innovation messages to attend to while they also have to make up their minds on which ones to disregard.

In the persuasion function, potential adopters must decide on the messages to seek out and the ones to ignore, whereas; in the decision function, the available options are different from what were obtainable in the previous stages. The decision stage presents with two possibilities which are either to accept or refuse a new idea. This decision involves an immediate consideration of having to try out an innovation or not to give it a try at all. Most farmers will accept an idea/technology only after trying it out; and it is after the trial that he can then determine its suitability.

Climate-Smart Agriculture (CSA)

CSA derives from an attribute in its terminology; that is, smart way of practicing agriculture. It is a form of agriculture that sustainably increases agricultural output and earnings; it enhances adaptation and aims to build resilience to the effect of climate change by reducing or eliminating Greenhouse Gases (GHGs) wherever possible, as well as advancing the attainment of global food safety (FAO 2014). CSA is mainly targeted at integrating the three broad areas of sustainable development which are the economic, social, and environmental components; by equally justifying food security, ecosystem management, and climate change problems.

Food security can be enhanced by maintaining an increasing agricultural production efficiency while the ecosystem can be more easily managed by adapting to, as well as building resilience to climate change. On the other hand, the climate change problems can largely be well mitigated by either working to lessen and/or eradicate greenhouse gases (GHG) emissions. However, CSA is never a pre-arranged condition, neither is it a given template that can be generally applied or practiced. Suffice to say that, it is a process that calls for site-specific assessments of the social, economic and ecological conditions to identify proper farming technologies that are appropriate in any given situation. It is important to note that a key component of CSA is the incorporated landscape method which suggests that the CSA practices that are being employed and/or applied in a given environment would follow the main beliefs of ecosystem management and sustainable land and water utilization. Climate-smart agriculture aims at supporting livelihoods at farm level by ensuring food security of smallholder farmers and helping to improve the management and utilization of natural resources. It also fosters the adoption of appropriate methods and skills for the production, processing and marketing of agricultural supplies. Whereas at the national level, CSA pursues legitimate and relevant technological and financial policies that would support nations to establish/entrench climate change adaptation into their agricultural sector. There is, at the national level, a considerable progressing efforts to promote CSA in Africa while ECOWAS also made efforts to support the mainstreaming of CSA into the ECOWAP/CAADP programs (ECOWAS 2015). In Nigeria specifically, there are research institutions and colleges owned by the Federal Government; who are working on CSA in order to render their own contribution to the fight against climate change related problems (FAO and ICRISAT 2019).

Benefits of Climate-Smart Agriculture

Climate-smart agriculture, as noted earlier, underscores the importance of building evidence to highlight practicable alternatives to the existing non-climate friendly approaches. It also encourages important enabling activities that can translate into better productivity and a friendlier environment. Through CSA, attention can be drawn to agricultural systems (crop, livestock, aquaculture, and agroforestry) that support the ecosystem services as this would enhance productivity, adaptation and mitigation; in a way that would encourage smart farming practices in order to reduce the effect of climate change (Adedeji et al. 2017). The CSA process offers stakeholders, such as farmers and other main players in the sector; the techniques and gadgets, as well as management and styles required to regulate more challenging and tougher production systems.

The FAO, who has immensely contributed to the development of CSA in Africa, also trained numerous individuals on the CSA approach and have equally developed a program called Safe Approach to Fuel and Energy (SAFE) in northeast Nigeria (FAO 2019). The SAFE program is aimed at reducing the need for firewood and by extension, the release of greenhouse gas is further reduced. Among the documented efforts at reducing the impact of climate change is the Oxfam Nigeria and the European Union collaboration which designed a project named Pro-Resilience-Action (PROACT). PROACT started in Nigeria in 2016 with enhancement of food security as its purpose while helping farmers build enduring means of sustenance (FAO and ICRISAT 2019a). The project has Mubi South, Fufore, Song, and Guyuk local government areas (LGAs) of Adamawa State currently participating where certain relevant and appropriate climate-smart actions are being combined to support famers become more resilient. Some of the actions, which are rolled out in phases, are provision of farm inputs and training of farmers' groups.

In the four locations of Adamawa State (Mubi South, Fufore, Song, and Guyuk), inputs such as fertilizers and water pumps are supplied to farmers while extension workers are being trained on rural investments and formation of loan groups in order to enable them access flexible financing services (FAO and ICRISAT 2019a). So far, about 700 farmer groups (women constitute more than half of the population) have been provided with farm inputs. The scheme has documented some level of success as rice growers have seen a boost of almost 100 percent despite the dry season rice production (Adedeji et al. 2017). This was made possible due to the distribution of inputs such as improved seeds, water pumps, among others. The project did not only focus on input provision for farmers as about 70 government extension workers in Kebbi and Adamawa States have received training on better environment-friendly agricultural practices for improved cereals production (Adedeji et al. 2017; FAO and ICRISAT 2019a). Farmer Training Fields (FTF) were then established wherein the extension workers in PROACT Communities assist farmers in order to boost their capacity as well as support them to adopt the best practices that can improve their yields and productivity. It has been observed through field evidences that the FTF program has contributed to higher acceptance of CSAPs and this has translated into enhanced food safety and income for farmers (FAO and ICRISAT 2019a).

In the short run, the project aims to handover improved farming skills to farmers as well as equip them to plant 500,000 trees to battle desertification and climate change; while in the long run, it is expected that about 1,400 community investments and loan groups would have been established. Another observable CSA effort is seen in adoption of terracing as a climate-smart practice in the southern and northern parts of Borno State as a result of the rocky topography of the area (FAO and ICRISAT 2019b). A farmer, while confirming the use of terracing (even without the knowledge or advent of CSA) for more than forty years, confirmed that it helped to combat the effects of soil erosion by preserving moisture and crop nutrients on the farm.

The farmer affirmed a higher yield of sorghum and maize on terraced farms (47% higher) when compared to non-terraced ones. Added to the advantage of terracing in the area is the assertion that farm events such as weeding and spraying are better managed on terraced farms while suggesting that farming in the entire southern part of Borno might have been very unprofitable if terracing was not being practiced (FAO and ICRISAT 2019b). Among the manifestation of the benefits of CSAPs is yet another evidence as provided by a farmer, in Potiskum LGA of Yobe State, who planted improved variety of millet (SOSAT) and got higher yields over and above the traditional variety (Gwagwa) that he had been planting previously (FAO and ICRISAT 2019c). Given the nature of local farmers, he initially resisted the new SOSAT variety as a replacement for his Gwagwa local variety. Hence, he intercropped SOSAT and Gwagwa millets with sorghum and cowpea as commonly practiced around Potiskum. This was done for 4 years while monitoring the variations in yield (FAO and ICRISAT 2019c). The farmer realized 550 kg/ha from the traditional millet while SOSAT millet produced 900 kg/ha in the first year. The following 2 years, on the same plots, but only intercropped with cowpea without sorghum, produced 350 kg/ha of SOSAT over the traditional variety. The first year reflected a marked increase of 64% yield over the utilization of the traditional local variety while the next year produced yield increase of 88%. This field experiment is yet another indication that climate-smart options, such as planting of improved varieties, could help enhance crop output and diversification which can also potentially increase profits of farmers (Tiamiyu et al. 2018b; FAO and ICRISAT 2019c).

Adoption of Climate-Smart Agricultural Practices by Rural Farmers in Nigeria

Smallholder farmers have, in the past, accidentally practiced CSA as part of their habitual farming system in Northern Nigeria (Fanen and Olalekan 2014). Similarly, it was also found in another study that the foremost strategy adopted by farmers to deal with climate change was to plant drought resistant varieties (Naswem et al. 2016). Other strategies used in mitigating the effect of climate change include: relocation from climate risk zones, prayers to God for divine intervention, multiple cropping, recycling of waste product, and improvement on farmers' management skills. According to Ojoko et al. (2017), the five most used CSAPs in Nigeria included conservation agriculture, application of organic manure, crop diversification, usage of wetland (Fadama), planting of drought tolerant crops, and agroforestry. Tiamiyu et al. (2017) in their study also reported similar findings where the adoption of CSAPs included climate smart approaches such as planting of drought tolerant and early maturing varieties, application of organic compost, use of cultural practices such as intercropping and crop rotation, composting rather than burning, and erection of terraces on sloppy/hilly farmland.

In other studies, Onyeneke et al. (2018) also reported that mixed farming, high yielding cultivars, agro-forestry, and other cultural practices such as mulching and zero tillage, as well as membership in development associations were the CSAPs adopted by the farmers in their study. Further evidence had showed that farmers in Nigeria have continued to adopt climate smart practices to mitigate the effect of climate change while simultaneously increasing productivity. For instance, female farmers adopted agro-forestry in derived savannah, guinea savannah, and rain forest, while the male farmers employed organic manure, zero tillage, and crop rotation in the same agro-ecological zones, respectively (Fapojuwo et al. 2018). However, it was observed that majority of farmers in the study never made use of the climate smart practices; hence low adoption rate because they were not even aware of the opportunities (Tiamiyu et al. 2018a). Nonetheless, planting of drought tolerant varieties and high yielding cultivars appear to be the most favored options accepted by farmers while only a few of the farmers used agro-forestry and integrated pest management as a smart way to curb the effect of climate change. Moreover, Adebayo and Ojogu (2019), in their study, had reported minimum or zero tillage, crop rotation, and other cultural practices such as mulching as the most favored CSAPs by farmers.

Challenges of Climate Smart Agriculture Practices

The adoption of CSAPs is linked to a myriad of impediments as there is an array of socio-economic and institutional complications. As usual, the need for significant upfront expenditures on the part of poor farmers, inadequate access to technical information that are potential solutions to farmers' problems, and the inability of farmers to convert the knowledge gained into practice; are part of the challenges associated with the adoption of CSAPs. Furthermore, some procedures that are climatically smart in nature and which are also associated to sustainable soil man-agement may not be compatible with farmers' customs; thereby leading to non-adoption as previous studies have shown that, the onward dissemination of ideas may greatly rely on knowledge in combination with social activities that farmers may not be capable of performing (World Bank 2018).

It had been reported that lack of funding portends a huge bottleneck in the acceptance of CSAP in Northern Nigeria while farmers also have little or no knowledge of CSAP; in addition to a poor understanding of the application of some of those practices by extension workers who would normally be the ones to disseminate technology to the famers. Other challenges include a dearth of stake-holders such as nongovernmental organizations; who can drive the campaign for CSAPs in the region (Fanen and Olalekan 2014). Moreover, while dealing with climate change, desertification poses a big threat for people, farmers inclusive. Hence, the subtle drive for tree planting by a few organizations makes the situation more problematic; thereby suggesting a neglect of the root cause. The implication of this is that there is a real need for a holistic and aggressive mobilization for tree planting. Meanwhile, limited opportunities for local managers to partake in policy formulation/decisions at international and local levels is a further weakness in the effort to combat climate change impact by encouraging adoption of CSAPs. Other hindrances restraining farmers from using CSAPs include: lack of government's support, high cost of executing the practices, and insufficient training on the prac-tices (Adebayo and Ojogu 2019). Specific challenges of CSA, according to Williams et al. (2015) are outlined:

1. Lack of practical understanding of the approach: The need for a full comprehension of protocols surrounding the use of CSAPs is important in order for messages to be clearly communicated to farmers. Thus, the effort for scaling up initiatives will depend on the understanding of the CSA idea by relevant stakeholders.

2. Inadequate data and information: It is always a herculean task for African nations both at national and local levels to access and utilize suitable tools for climatic and landscape level data.

3. Inadequate investment in CSA: Again, it has been repeatedly found that investments required to scale up adaptation and resilience in agriculture is lacking right from national up to farm levels. The Africa's Infrastructure Diagnostic Report (Foster and Briceño-Garmendia 2010) also reported a deficit in social infrastructure such as road, transportation, communication, power and water; which are high requirements for the uptake of CSAP.

4. Socio-economic constraints: Access to productive resources like land, transportation and water, as well as human resources were the limitations at the farm level.

5. Inadequate women empowerment: Women who contribute significantly to food production in Africa are largely marginalized when it comes to access to and control over productive resources. For instance, it was found in a previous study carried out in northern Nigeria that women do not own and control productive assets such as land, crop inputs, livestock inputs, and capital (Aderinoye-Abdulwahab et al. 2016) and that this makes them to be poorly empowered. It is noteworthy that such gender stereotypes relating to access to land rights, education, technologies, and credit; are among the countless stumbling blocks to women's effective contribution in the farming sector.

6. Poor funding and efficient risk-sharing schemes: Funding devices to support the uptake of CSAP are not yet popular in Africa nations. Whereas, huge investment is required for the transition to climate-smart agricultural development pathways as farmers continue to face the risk of accepting new technologies. Meanwhile, the benefits of the technologies often only come after several years/seasons of production, thus the realization of a wider coverage of CSAPs becomes problematic.

7. Difficulty in running trade-offs from the farmers' and policymakers' perspectives: The dissimilarity in objectives between farmer groups and policy makers continues to cause a difference of opinion in respect of priorities for resource management.

Conclusion and Recommendation

The adoption of smart agricultural practice is an integral part for increasing agricultural productivity as it enhances adaptation to climate change. It also helps to reduce greenhouse gases emissions while enhancing the attainment of food security. The popular CSAPs adopted by rural farmers in Nigeria are: conservation agriculture, adoption of organic compost, use of wetland (Fadama), utilization of drought tolerant crops, relocation from climate risk zones, and use of cultural and management practices such as mulching and prayers for divine intervention. The challenges militating against the adoption of CSAPs by farmers in the study include: poor funding, lack of practical understanding of the approach, inadequate access to information on CSAPs, and socio-economic constraints at the farm level.

It would, hence, be recommended that there is a dire need for an aggressive awareness and mobilization campaign for the use of CSAPs as a solution to climate change impact in order to increase farmers' productivity. Additionally, this chapter would also recommend that government and other stakeholders should ensure that productive resources such as land, planting technologies, education and other social amenities be made available and accessible to farmers generally, but more especially women; in order to facilitate easy uptake of climate smart practices. Moreover, there is a need to motivate and encourage farmers to adopt climate smart practices, and one way of achieving this is by incorporating funding devices to support CSAPs and other climate friendly support mechanisms in policy documents.

References

Adebayo AE, Ojogu EO (2019) Assessment of the use of climate smart agricultural practices among smallholder farmers in Ogun state. Acta Sci Agric 3(6):47–56

Adedeji IA, Tiku NE, Waziri-Ugwu PR, Sanusi SO (2017) The effect of climate change on rice production in Adamawa State, Nigeria. *Agroeconomia Croatica* 7(1):1–13. ISSN 1333-2422

Adekoya AE, Tologbonse EB (2011) Adoption and diffusion of innovations; agricultural extension in Nigeria, 2nd edn. Agric Exten Soc Niger, Ilorin, pp 36–48

Aderinoye-Abdulwahab SA, Dolapo AT, Matanmi BM, Adisa RS (2016) Ownership of productive resources: a panacea for empowering rural women in Kwara state, Nigeria. J Prod Agric Technol 12(1):17–26

African Climate Smart Agriculture Summit (2014) Twenty-third ordinary session. Equitorial Guinea, Malabo. 26–27 June 2014. http://africacsa.org/introducing-the-africa-csa-alliance-ascaa/

ECOWAS (2015) A Thematic Roundup of Research, decision-making support and capitalization-extension of endogenous and scientific knowledge for the CSA in the ECOWAS/UEMOA/CILSS region. Paper prepared for the High Level Forum of Climate-Smart Agriculture (CSA) Stakeholders in West Africa, Bamako, 15–18 June 2015

Ekong EE (2010) An introduction and analysis of rural Nigeria, 3rd edn. Dove Educational Publishers, Uyo, pp 280–266

Fanen T, Olalekan A (2014) Assessing the role of climate-smart agriculture in combating climate change, desertification and improving rural livelihood in Northern Nigeria. Afr J Agric Res 9 (15):1180–1191

FAO (2014) Innovation in family farming. The State of Food and Agriculture Series. A production of Food and Agricultural Organization of the United Nations, Rome. E-ISBN 978-92-5-108537-0. Available at: http://www.fao.org/3/a-i4040e.pdf. 161p

FAO (2019). http://www.fao.org/faostat/en/#data/GT

FAO, ICRISAT (2019a) Climate-smart agriculture in the Adamawa state of Nigeria. CSA Country Profiles for Africa Series. International Center for Tropical Agriculture (CIAT); International Crops Research Institute for the Semi-Arid Tropics (ICRISAT); Food and Agriculture Organization of the United Nations (FAO), Rome. 22p

FAO, ICRISAT (2019b) Climate-smart agriculture in the Borno state of Nigeria. CSA Country Profiles for Africa Series. International Center for Tropical Agriculture (CIAT); International Crops Research Institute for the Semi-Arid Tropics (ICRISAT); Food and Agriculture Organization of the United Nations (FAO), Rome. 22p

FAO, ICRISAT (2019c) Climate-smart agriculture in the Yobe state of Nigeria. CSA Country Profiles for Africa Series. International Center for Tropical Agriculture (CIAT); International Crops Research Institute for the Semi-Arid Tropics (ICRISAT); Food and Agriculture Organization of the United Nations (FAO), Rome. 22p

Fapojuwo OE, Ogunnaike MG, Shittu AM, Kehinde MO, Oyawole FP (2018) Gender gaps and adoption of climate smart practices among cereal farm households in Nigeria. Niger J Agric Econ 8(1):38–49

Foster V, Briceño-Garmendia C (2010) Africa's infrastructure. A time for transformation. A copublication of the Agence Française de Développement and the World Bank. World Bank, Washington, DC

Ifeanyi-obi CC, Nnadi FN (2014) Climate change adaptation measures used by farmers in South-South Nigeria. J Environ Sci Toxicol Food Technol 8(4):1–6. https://pdfs.semanticscholar.org/a63d/22bf8b8ebde892a7c9d761ee1653e0e11df6.pdf

IPCC (2012) Managing the risks of extreme events and disasters to advance climate change adaptation. a special report of working groups I and II of the Intergovernmental Panel on Climate Change (eds: Field CB, Barros V, Stocker TF, Qin D, Dokken DJ, Ebi KL, Mastrandrea MD, Mach KJ, Plattner G-K, Allen SK, Tignor M, Midgley PM). Cambridge University Press, Cambridge/New York, p 582

Naswem AA, Akpehe GA, Awuaga MN (2016) Adaptation strategies to climate change among Rice farmers in Katsina-Ala local government area of Benue state, Nigeria. IOSR J Agric Veter Sci 9 (10):33–37

Ojoko EA, Akinwunmi JA, Yusuf SA, Oni OA (2017) Factors influencing the level of use of climate-smart agricultural practices (CSAPs) in Sokoto state, Nigeria. J Agric Sci 62(3):315–327

Olorunfemi TO, Olorunfemi OD, Oladele OI (2019) Determinants of the involvement of extension agents in disseminating climate smart agricultural initiatives: implication for scaling up. J Saudi Soc Agric Sci. https://doi.org/10.1016/j.jssas.2019.03.0031-8

Onyeneke RU, Igberi CO, Uwadoka CO, Ogbeni JA (2018) Status of climate-smart agriculture in Southeast Nigeria. GeoJournal 83:333–346. https://doi.org/10.1007/s10708-017-9773-z

Ray GL (2011) Extension communication and management, 8th edn. Kalyani Publishers, New Delhi, pp 147–170

Rogers E (2003) The diffusion of innovations. (5th ed.). New York, NY: The Free Press

Tiamiyu SA, Ugalahi UB, Fabunmi T, Sanusi RO, Fapojuwo EO, Shittu AM (2017) Analysis of farmers' adoption of climate smart agricultural practices in Northern Nigeria. In: Proceedings of the 4th International Conference on Agriculture and Forestry, vol 3, pp 19–26

Tiamiyu SA, Ugalahi UB, Eze JN, Shittu MA (2018a) Adoption of climate smart agricultural practices and farmers' willingness to accept incentives in Nigeria. Int J Agric Environ Res 4 (4):198–205

Tiamiyu SA, Ugalahi UB, Fabunmi T, Sanusi R, Fapojuwo E, Shittu A (2018b) Analysis of farmers' adoption of climate smart agricultural practices in Northern Nigeria, pp 19–26. https://doi.org/10.17501/icoaf.2017.3104

USAID (2016) Climate smart agriculture in feed the future programs. Feed THE Future (The U.S. Government's Global Hunger & Food Security Initiative). Available at: https://www.climatelinks.org/sites/default/files/asset/document/Framework%20CSA%20paper%20final%20%281%29.pdf. 13p

Williams TO, Mul M, Cofie O, Kinyangi J, Zougmore R, Wamukoya G, Nyasimi M, Mapfumo P, Speranza CI, Amwata D, Frid-Nielsen S, Partey S, Girvetz E, Rosenstock T, Campbell B (2015) Climate smart agriculture in the African context. Abdou Dlouf international conference center, Feeding Africa an Action Plan for African Agricultural Transformation

World Bank (2010) Development and climate change. World Development report 2010. World Bank, Washington, DC

World Bank (2018) Scaling up climate-smart agriculture through the Africa climate business plan. World Bank, Washington, DC

8

Improving Food Security by Adapting and Mitigating Climate Change-Induced Crop Pest: The Novelty of Plant-Organic Sludge in Southern Nigeria

Chukwudi Nwaogu

Contents

C. Nwaogu (✉)
Department of Forest Protection and Entomology, Faculty of Forestry and Wood Sciences,
Czech University of Life Sciences, Prague 6-Suchdol, Czech Republic

Department of Environmental Management, Federal University of Technology, Owerri, Nigeria

Department of Ecology, Faculty of Environmental Sciences, Czech University of Life Sciences,
Prague 6-Suchdol, Czech Republic
e-mail: cnwaogu@gmail.com

Abstract

Climate change is a global issue threatening food security, environmental safety, and human health in tropical and developing countries where people depend mainly on agriculture for their livelihood. Nigeria ranks among the top in the global yam production. It has the largest population in Africa and has been able to secure food for its growing population through food crops especially yam. Unfortunately, the recent increase in termites' colonies due to climate change threatens yam yield. Besides harming man and environment, pesticides are expensive and not easily accessible to control the pests. This prompted a study which aimed at applying a biotrado-cultural approach in controlling the termites, as well as improving soil chemical properties and yam production. The study hypothesized that *Chromolaena odorata* and *Elaeis guineensis* sludge improved soil nutrient and yam yield and consequently decreased termites' outbreak. In a randomized design experiment of five blocks and five replicates, five different treatments including unmanaged (UM), *Vernonia amygdalina* (VA), *Chromolaena odorata* (CO), *Elaeis guineensis* (EG) liquid sludge, and fipronil (FP) were applied in termites-infested agricultural soil. Data were collected and measured on the responses of soil chemical properties, termites, and yam yield to treatments using one-way ANOVA, regression, and multivariate analyses. The result showed that *Chromolaena odorata* (CO) and EG treatments were the best treatments for controlling termites and increase yam production. Termites were successfully controlled in VA and FP treatments, but the control was not commensurate with yam production. The experiment needs to be extended to other locations in the study region. It also requires an intensive and long-term investigation in order to thoroughly understand (i) the influence of climate change on the termites' outbreak, (ii) the extent of termite damage to the crops, (iii) the impacts of climate change and variability on yam yields, (iii) the agricultural and economic benefits of the applied treatments, and (iv) the ecological and human health safety of the treatments.

Keywords

Climate change · Sustainable agriculture · Soil · Pest management · Termites · *Elaeis guineensis* · *Vernonia amygdalina* · *Chromolaena odorata* · Fipronil · Ikpo-Obibi

Introduction

Climate change has globally become a serious threat to environment and man especially in the areas of food security and rapid growing population. Though the impacts of climate change have no geographical boundary, yet the countries in sub-Saharan Africa tend to suffer more because of several reasons including socio-economic, political, and ecological factors (Slingo et al. 2005; Kurukulasuriya et al.

2006). The Assessment Reports of the Intergovernmental Panel on Climate Change (IPCC) urged that, even with the predicted climate change scenarios, extreme events may still occur with devastating effects in more vulnerable areas, causing severe long-term food insecurity (Boko et al. 2007; Christensen et al. 2007).

Yam (*Dioscorea* spp.) is a tuber crop which serves as a major staple food for about 34% of the world's population. In comparison to other tuber crops, yam is a vital source of essential minerals including carbohydrates, vitamins, proteins, and dietary fibers (Olajumoke et al. 2012). In 2004, the world's yam production was estimated at about 46.8 million metric tons, and West Africa accounted for more than two-third of this global production (Sartie et al. 2012). Nigeria ranks highest in yam production among the sub-Saharan African countries (CGIAR 2004). Yam is a highly preferred food crop in Nigeria because:

(i)   It is one of the easiest and fastest food to be prepared in different flavor.
(ii) Yam tubers can be preserved for longer time (4–6 months) at ambient temperature unlike sweet potato (*Ipomoea batatas* L.) and cassava (*Manihot esculenta*). The sustainability of yam as a source of food for every household is high, even during the onset of the rainy season when food tends to be scarce, yam tubers will be available (Loko et al. 2015). Hence, yam is locally referred as "enyi nwa-ogbuenye na'uwu" meaning "the orphan's salvager in time of famine."
(iii) Economically, yam tubers are high sources of income for the indigenous farmers, and this helps to alleviate poverty (Olorede and Alabi 2013).
(iv) Socio-culturally, yam promotes the social life of the people: several festivals (such as New Yam Festival, Yam title coronation, and marriage) are celebrated by communities, groups, and individuals using yam (Osunde and Orhevba 2009).

During cultivation, yam has its tubers buried in the soil which is the habitat of most ant species especially termite. Termite (*Isoptera*) has been described as one of the yam tubers' main destructive fauna (Atu 1993; Loko et al. 2015). Common among the termite species that damage the yams are the *Microtermes*, *Ancistrotermes*, and *Macrotermes* (Loko et al. 2015). The damage consists of feeding and destruction of planted setts of yam (including tubers, leaves, stems), and yam staking materials, as well as release of methane (Zimmerman and Greenberg 1983). Yam tubers to be harvested are sometimes heavily tunneled in termite-infested soil because the *Microtermes* spp. seldom build colonies within tubers and create apparent hollows. In Nigeria, it has been reported that in soils with high termite infestation, farmers lose more than 5 t ha^{-1} of yam due to damage by termites (Atu 1993). However, in the tropics and subtropics, termite has been reported as litter decomposer, and a key player in soil formation (Richard et al. 2006; Belyaeva and Tiunov 2010), yet it degrades the soil quality (Devendra et al. 1998).

Globally, the net impact of climate change has been predicted to include increase in pest damage to agricultural resources (Lovett et al. 2005). In tropical Africa precisely, climate change has crucial role in the striving, reproduction, and crop-destructive ability of the pests by making the environment favorable for them. Due to variations in weather especially rainfall and temperature, termites' outbreak and infestation become acute. Studies have shown that an average annual rainfall and temperature below 100 cm and 27°C, respectively, promote the survival and catastrophic impacts of the pest (Atu 1993; Richard et al. 2006; Belyaeva and Tiunov 2010). Though termites have been reported to be present throughout the year in most tropical countries of Africa (Wood 1995), their agricultural and economic damage tend to be exacerbated during periods of low rainfall (Ahmed et al. 2011). In the study conducted on the potential impact of climate change on termite distribution in Southern Zambia, Ahmed et al. (2011) found that after a drought, the number of pestiferous termite species increased drastically. Similarly, another study, performed in Uganda by Pomeroy (1976), demonstrated that the distribution of termites' mounds was significantly correlated with temperature and that large termite mounds were absent in areas of lower temperature and high rainfall. Kemp (1955) observed that climate was the principal factor determining the distribution of termites in Northeastern Tanganyika (Tanzania). Farmers in Nigeria, Uganda, and Zambia have reiterated that termite problems are more serious now than in the past (Atu 1993; Sekamatte and Okwakol 2007; Sileshi et al. 2009). Damage by the pests is higher during dry periods than in periods of regular rainfall (Logan et al. 1990). The rapid increase in termite damage might also be attributed to climate change-induced drought. For example, in the last two to three decades, drought associated with El Niño episodes has become more intense and widespread in Africa (Harrington and Stork 1995). Many studies have previously established strong nexus among climate change, termites invasion, and damage to agricultural resources (Jones 1990; Kemp 1955; Logan et al. 1990; Nkunika 1998; Bignell and Eggleton 2000; Eggleton et al. 2002; Ahmed and French 2008; Ahmed et al. 2011; Beaudrot et al. 2011; Rouland-Lefèvre 2011; Buczkowski and Bertelsmeier 2017).

Ikpo-Obibi is one of the important yam-producing communities in Nigeria with more than 80% of the population engaging in agriculture. It is pathetic to report the farmers' ordeal with the climate change-induced termites in the process of yam production. Despite the nutritional and socioeconomic benefits of yam, the termites' attack poses great challenge.

In terms of the control measures, most farmers in Africa have attempted the application of different pesticides, yet sustainable solution was never achieved. For example, in Nigeria, some farmers have applied pesticides such as fipronil, aldrin, imidacloprid, chlorpyrifos, sulfluramid, and heptachlor to control the pests by dressing the yam setts and farmlands. The problems of inaccessibility, high costs, human health safety, and ecological effects of these pesticides limit their usage (UNEP 2000; Boonyatumanond et al. 2006; Sánchez-Bayo 2014). The need for a sustainable method of ameliorating this yam pest and to have increased yield became crucial due to higher food demand from the growing population. In other West African country such as Ghana, it was reported that traditional methods (wood ash, cow dung, and aqueous extract of plant residues "dawadawa") have been successfully used to control the yam pests (Asante et al. 2008). The records about such traditional applications are yet to be found in Nigeria. Therefore, the present study aimed at appraising a new biotrado-cultural approach of coping with the climate change by controlling the termites, improving the soil properties, and increasing yam production using aqueous extract of *E. guineensis*, *V. amygdalina*, and *C. odorata*. The study also compared the impacts of these plant materials on the soil, yam yield, and termites in relation to the result from the fipronil treatment. It is hypothesized that (i) *C. odorata* enriched the soil minerals and increased yam yield by reducing the effects from climate change and the termites; (ii) the attraction of *E. guineensis* sludge to termites makes it an intervening material that reduced the pests' attack on the yam tubers and consequently enhanced the soil organic carbon (Corg); and (iii) *V. amygdalina* and fipronil treatments can successfully control termites but they have high concentrations of trace elements which aggravate the impacts of climate change and exert negative effects on the soil and yam production. Within this context, the following questions were addressed: (a) What are the variations in the concentrations of soil chemical properties, under the different treatments applied, and how significant are these concentrations? (b) Which of the applied treatments and season increased soil essential minerals? (c) Does increase in Corg and rainfall increase yam production? (d) How sustainable is the fipronil application in relation to termites' control, soil fertility, and yam yield? (e) To what extent does the variability in climate influence the termites and their impacts on yam production?

Materials and Methods

Study Area

Ikpo is one of the oldest villages among the eight villages in Obibi community in Etche Local Government Area (ELGA) of Rivers State, Nigeria. The farmland is located within latitude 5°06'52.7″N and longitude 7°11'59.4″E (Fig. 1) with an area of about 52.2 km^2 and a gentle sloping altitude ranging from 50 m to 100 m above sea level. The mean annual rainfall ranges from 100.4 cm to 241.7 cm and mean annual temperature ranging from 26.5 °C to 28.3 °C (SPDC 1998) (Fig. 2). The study was conducted from 2013 to 2016 under rainfed conditions. The mean monthly rainfall (cm) and temperature (°C) of the study site varied (Fig. 2) with 2013 and 2014 indicating normal rainfall years, 2015 revealing a dry year, and 2016 wet year. Though the study area has variations in rainfall that shows a bimodal rainfall distributional trend, yet no month without rainfall. Early-season rain begins in either late January or February and ends in July, and this is categorized as the early growing season. On the other hand, the late-season rain starts in August and ends in December (Nwaogu et al. 2017).

Geologically, the study site lies within Niger Delta Basin, and it is characterized by the Benin or coastal plain sand formation. The age of these formations ranges from Miocene to Eocene.

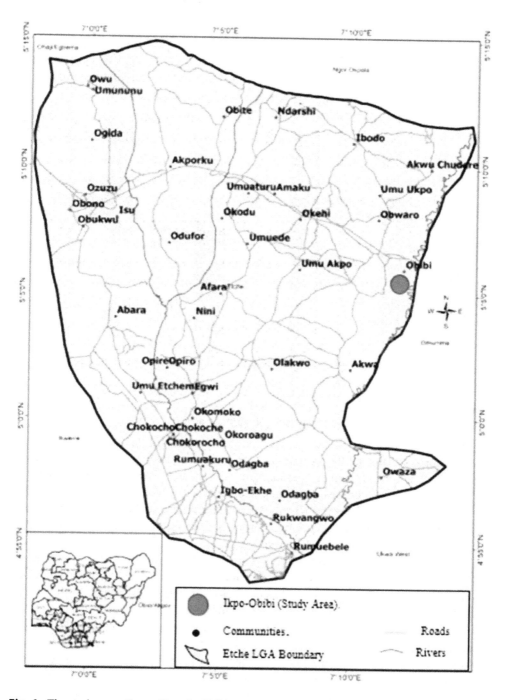

Fig. 1 The study area: Ikpo village in Obibi community, Etche LGA of Rivers State, Nigeria

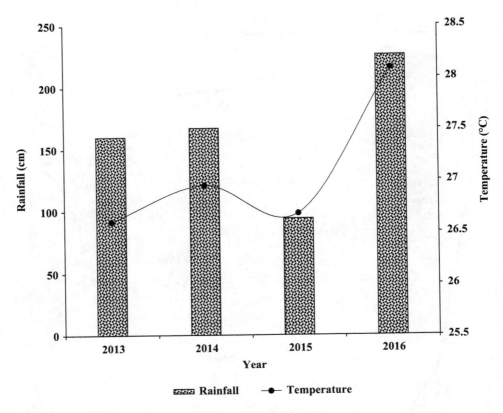

Fig. 2 Mean annual rainfall and temperature of the study area

The shallow parts of the formations are composed mainly of non-marine sand deposited in alluvial or upper coastal plain (Doust and Omatsola 1990). The soil composition of the study area consisted of fine alluvial content which is an extension of the rich sediments from the River Niger Delta alluvial soil. Presently, the heavy anthropogenic activities (e.g., mining, farming, grazing), pests, and environmental factors have influenced the soil properties (Table 1).

The study area is a typical rainforest region with traces of mangrove and freshwater swamp forests toward the south (Nwankwoala and Nwaogu 2009). There are bushes with trees and shrubs, as well as patches of grasslands dominated by elephant grasses where most farming activities are performed.

There are several termites' mounds which are between 2–7 m high and 2–5 m in diameter that cover the land space forming high termite mound density per hectare. Though the study site has for some years been dominated by termites, the activities of these pests to a large extent depend on climatic conditions. From the growing period to the maturity stage, the termites destroy the development of the crops planted at the site. This at most times leaves the rural farmers with little or nothing to be harvested. At planting and harvesting, five different species of termites were identified as the most common yam-destroying termites: *Amitermes evuncifer*, *Macrotermes bellicosus*, *Microtermes obesi*, *Trinervitermes oeconomus*, and *Trinervitermes geminatus* (Atu 1993; Loko et al. 2015).

Table 1 Soil (0–15 cm depth) chemical and physical properties in the study site prior to the experiment

Soil property	Mean ± standard error
Corg (mg kg^{-1})	5481 ± 311
Ntot (mg kg^{-1})	1197 ± 95
P (mg kg^{-1})	92.6 ± 23.3
K (mg kg^{-1})	471.1 ± 37.7
Mg (mg kg^{-1})	334.6 ± 15.9
Ca (mg kg^{-1})	421.4 ± 33.4
Fe (mg kg^{-1})	7.8 ± 1.5
Mn (mg kg^{-1})	6.1 ± 0.9
Cu (mg kg^{-1})	5.9 ± 1.2
Zn (mg kg^{-1})	8.8 ± 1.1
Cr (mg kg^{-1})	5.0 ± 0.3
pH	6.1 ± 0.5
Sand (g kg^{-1})	716 ± 10.4
Silt (g kg^{-1})	118 ± 6.1
Clay (g kg^{-1})	161 ± 4.3
Textural class	Sandy-loam
Bulk density (Mg m^{-3})	1.3 ± 0.1

Experimental Design, Yam Cultivation, and Treatments

The study was conducted in a randomized block design (Fig. 3). The area covered 5,400 m^2 with about 80% of the entire field being used for the yam cultivation and treatments, while the remaining percentage consisted of the borders around the entire treatment plots. Each plot consisted of 144 m^2 (12 m × 12 m) and was separated by 4 m (column) and 2 m (row) buffer zones. Land preparation for the experiment took place in 2012, and the first phase of the experiment was performed in 4 years (2013–2016) characterized with differences in annual rainfall and temperature. For instance, 2013 and 2014 had a normal rainfall, 2015 was a dry year, and 2016 recorded excess rainfall.

The experimental design includes five blocks, five replicates, five different treatments (UM, unmanaged; VA, *Vernonia amygdalina*; CO, *Chromolaena odorata*; EG, *Elaeis guineensis* liquid sludge; FP, fipronil) (Fig. 3), and their chemical compositions (Table 2).

The land was cleared and the plants' litter removed. Yam planting heaps of 80–100 cm high and 100–120 cm in diameter were prepared every year before the first rainfall of the year. After the first 2–3 rains of the year, yam setts were planted on the prepared heaps/mounds which were at the intervals of 15–20 cm with the cut face placed up (Fig. 4).

Based on the local tradition, the mounds were preferable to the ridges because they promoted easy staking and enhance the production of larger yam tubers when compared with ridges. The VA, CO, and EG treatments were traditionally processed by crushing, soaking in water, and squeezing out the sludge which were stored in 0.7 liter plastics (Appendix Figs. 8, 9, 10, 11, and 12). These were preserved in a cool

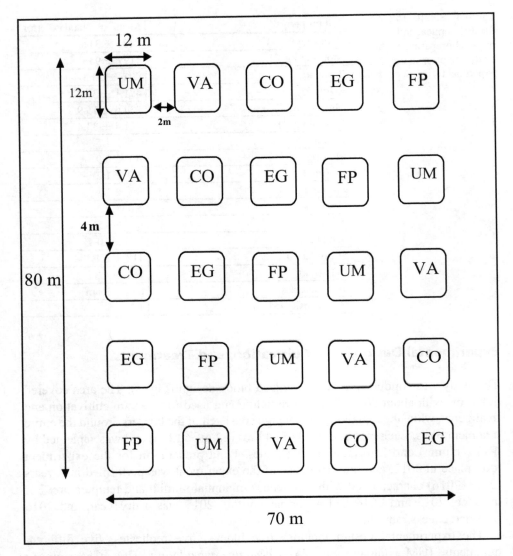

Fig. 3 Experimental design block for the study. Treatments were UM (unmanaged), VA (*Vernonia amygdalina*), CO (*Chromolaena odorata*), EG (*Elaeis guineensis* liquid sludge), FP (fipronil)

temperature to prevent fermentation before being applied to the yam heaps. On the other hand, the aqueous dilution of fipronil (Fipronil (120068-37-3); Alina Zhao, Ningbo Samreal Chemical Co., Ltd., China) was prepared as prescribed on the product label for termites' control. The prepared treatments were applied by dressing the yam setts. Between 0.2 and 0.3 litters of sludge (or aqueous) were dropped in and around every heap where the yam was planted. This was performed three times (in February/March, August, and November) for each growing season. By using hoes, manual weeding was performed three times (in late May, August, and November) for each growing season. Two stakes, each of 200–300 cm in height,

Table 2 Mineral and trace element contents (mg L^{-1}) of the organic materials used for the treatments (mean ± standard error of the mean)

		Treatment material		
		VA	CO	EG
Minerals	P	83.9 ± 5.6	116.1 ± 23.5	154.6 ± 12.1
	K	52.1 ± 7.3	181.9 ± 15.2	271.2 ± 25.6
	Mg	33.8 ± 2.9	104.6 ± 27.7	67.6 ± 7.3
	Ca	76.4 ± 11.1	93.3 ± 12.9	48.7 ± 2.5
Trace elements	Fe	36.1 ± 9.7	22.8 ± 8.1	11.3 ± 1.0
	Mn	3.3 ± 0.4	0.4 ± 0.0	0.3 ± 0.0
	Cu	5.2 ± 1.6	0.7 ± 0.1	0.9 ± 0.2
	Zn	2.4 ± 0.7	0.2 ± 0.0	0.8 ± 0.3
	Cr	2.3 ± 0.3	0.1 ± 0.0	0.2 ± 0.0

Fig. 4 Experiment site with yam soil heaps during planting and one of the termites' mounds

were used for staking the yam plants to vine over them – one stake for two plants and the other stake used for bracing the adjacent. Yam tuber harvesting was done in December each year following the local tradition.

Soil Sampling and Chemical Properties Analyses

In February/March and December, five soil sub-samples (0–15 cm soil depth) from individual treatment plot were randomly collected using a graduated auger. The soil samples were mixed and air-dried; visible pebbles, biomass residues, roots, and other organic debris were removed. Before taken to the laboratory for analysis, the samples were ground in a mortar to pass a 2 mm sieve. All measurements and analyses were performed within 30 days of sampling. The soil pH (H$_2$O) was determined by the method of McLean (1982). Organic carbon (Corg) was measured using the Walkley and Black wet oxidation method (Nelson and Sommers 1996). Total nitrogen (Ntot) was analyzed using micro-Kjeldahl method (Bremner and Mulvaney 1982). Available P, K, Ca, and Mg were extracted using Mehlich III solution (Mehlich 1984) with reagent and concentrations determined by introducing the inductively coupled plasma-optical emission spectrometry (ICP-OES: 720 Series, Agilent Technologies, USA). On the other hand, the concentrations of the trace elements (Cu, Mn, Cr, Fe, Zn) were determined following the Clayton and Tiller (1979) method. The concentrations were further analyzed using ICP-OES analyzer. The mean of five sub-samples from each monitored treatment was used for the statistical analyses.

Termites Sampling

The number of termites per heap was collected three times per year: in February/ March (during planting), August (during the short break, that is, the August rainfall break), and December (during harvesting). These periods coincided with the low rainfall months which formed the peak season for the termites' activities (Loko et al. 2015). The termites were observed and counted using the Zoom Handheld Lighted Magnifier Glass (MagniPros 5.5″, USA) by randomly selecting the yam heaps and excavating through to the soil profile where the yam tubers could be found. One-third of the heaps were sampled per plot.

Yam Yields

At physiological maturity, after 9–10 months (that is, in December), the yam tubers were harvested by randomly selecting the yam heaps and excavating through to the soil layers where the yam tubers could be found. One-third of the heaps were sampled per plot. The harvested tubers were washed using water. They were thereafter weighed, and result was recorded in yield per hectares.

Statistical Analyses

To test the effects of treatments on the soil chemical properties, yam yield, and number of termites, a one-way ANOVA was used. In addition, a repeated-measures ANOVA was applied to deduce the effect of year, treatment, and year x treatment interaction for number of termites, yam yields, soil minerals, and trace elements. The ANOVA was applied after the assumptions of normality were met. Regression analysis was used to test the relationship between yam tuber yield (t ha^{-1}) and number of termites per yam soil heap. All analyses were conducted using the IBM SPSS Statistics Version 20 (IBM Corporation 2011) (www.ibm-spss-statistics. soft32.com) and the STATISTICA 13.0 software (StatSoft, Tulsa, OK, USA), while all data were expressed as means of five replicates.

Furthermore, a multivariate analysis such as redundancy analysis (RDA) followed by a Monte Carlo permutation test with 999 permutations in the Canoco 5.0 software (Šmilauer and Lepš 2014) was used to measure the effect of the different treatments on soil chemical properties during the 4 years. Ordination diagram was produced by using the CanoDraw program software which prompted the presentation and visualization of the RDA results of the experiment.

Results

Soil Chemical Properties

The concentrations of all the monitored soil chemical properties were significant under the different treatments except for Mn and Cu (Table 3). Most of the measured minerals showed higher concentrations under the CO treatment when compared with VA, EG, UM, and FP treatments. On the other hand, VA and FP treatments had higher trace element concentrations and higher pH when compared with the other treatments. All the essential soil minerals were significantly affected by time and treatment, but not all were affected by the combination of both (i.e., time × treatment interaction) (Table 4).

No significant effect of treatment on soil chemical properties calculated by RDA was recorded in 2015 (Table 5). The variability of soil chemical properties explained by treatments subsequently increased from the initially 24% in 2013 to more than 25% in 2014 and more than 50% in 2016. Variability explained by treatments was constantly above 20% between 2013 and 2016.

The treatments were categorized into three different groups in relation to the concentrations of soil chemical properties as were revealed by the ordination diagram (Fig. 5). Based on the RDA analysis of the data collected during the 4 years, CO and EG treatments formed the first group, VA and FP treatments as the second group, and UM treatment as the third group. The relationships between individual soil element and treatment are visible from the ordination diagram (Fig. 5).

Yam Tuber Yield and Termites

The mean annual yam tuber yields under CO and EG treatments were significant, while UM, VA, and FP treatments never showed any significant differences during the study years (Fig. 6). Year 2015 had the lowest yam tuber yields under the different treatments, while 2013 had the highest.

The results from the total number of termites per yam heap showed that UM and EG treatments were significant at $P < 0.05$, while records for VA, CO, and FP treatments revealed no statistical differences in the years (Fig. 7). Year 2015 had the highest number of termites per heap, whereas 2016 had the lowest. The relationships between yam tuber yield and number of termites per heap were negatively significant under the CO and EG, while UM treatment ($R^2 = 0.50$; $P = 0.045$) showed a marginal weak relationship (Table 6).

Table 3 Mean concentrations (mg kg^{-1}) of soil chemical properties (minerals and trace elements) under the different treatments in 2013–2016. *P*-value represents corresponding probability value. Numbers represent the average of five replicates; ± represents standard error of the mean (SEM); significance differences ($P < 0.05$) between treatments in accordance with the Tukey's post hoc test are shown by different letters in the row (a < b < c < d < e). Treatment abbreviations (UM, VA, CO, EG, and FP) are described in Fig. 3

	UM	VA	Treatments CO	EG	FP	F-ratio	P-value
Minerals							
Corg	5326 ± 106ab	4881 ± 211a	7087 ± 462d	6139 ± 304c	5941 ± 84b	21.2	<0.001
Ntot	897.5 ± 114a	906.6 ± 125a	1815 ± 93c	1699 ± 281b	923 ± 75a	8.5	0.039
P	71.7 ± 23.5a	98.2 ± 16.1ab	203.6 ± 19.9c	117.9 ± 15.32b	95.8 ± 11.7ab	0.7	<0.001
K	221.9 ± 12.2ab	205.1 ± 5.7a	328.4 ± 45.3c	239.2 ± 29.6b	228.0 ± 15.8ab	5.3	0.027
Mg	341.8 ± 75.2c	287.9 ± 73.6b	469.6 ± 66.5e	413.3 ± 81.1d	116.3±28.4a	2.6	< 0.001
Ca	432.2 ± 59.6c	311.4 ± 54.2b	791.8 ± 83.7e	609.6 ± 61.4d	207.1 ± 19.9a	17.1	<0.001
Trace elements							
Fe	8.1 ± 1.5a	41.8 ± 7.3c	24.6 ± 4.1b	13.0 ± 1.5ab	52.3 ± 6.1d	9.1	0.022
Mn	3.1 ± 0.1	3.9 ± 0.7	3.4 ± 0.3	3.8 ± 0.2	4.2 ± 0.5	0.4	0.621
Cu	5.3±0.5	5.1±0.9	5.5±1.1	5.6±1.0	6.7±1.1	1.8	0.948
Zn	8.3±1.1b	17.4±2.5d	6.8 ± 2.1a	10.3 ± 2.3c	27 ± 3.8e	0.9	<0.001
Cr	5.7 ± 0.3a	25.9 ± 2.0c	7.1 ± 1.4ab	9.7 ± 1.8b	32.1 ± 3.1d	3.6	0.031
pH	6.0 ± 0.8a	7.5 ± 0.2d	6.4 ± 1.1b	6.9 ± 0.7c	8.1 ± 0.8e	0.2	0.043

Table 4 Results of repeated-measures ANOVA (time, treatment, time × treatment) of soil chemical properties, number of termites, and yam yields. df, degree of freedom; F, value derived from F-statistics in repeated-measures ANOVA; and P, probability value

| | Effect | | | | | |
| | Time: df = 4 | | | Treatment: df = 5 | Time × treatment: df = 20 | |
	F-ratio	P-value	F-ratio	P-value	F-ratio	P-value
Soil chemical properties						
Corg	23.5	*	11.8	*	3.1	*
Ntot	14.1	*	24.3	*	1.8	*
P	3.9	*	5.9	*	0.3	ns
K	3.3	*	17.7	*	0.5	ns
Mg	10.2	*	6.8	*	2.4	*
Ca	9.7	*	18.3	*	1.2	*
Fe	0.6	ns	35.1	*	0.4	ns
Mn	0.9	ns	1.8	ns	0.7	ns
Cu	1.8	ns	0.7	ns	0.5	ns
Zn	31.7	*	11.2	*	2.7	*
Cr	25.5	*	57.1	*	3.1	*
pH	14.1	*	8.5	*	0.5	ns
Number of termites	10.6	*	17.9	*	0.9	ns
Yam yield	37.3	*	22.3	*	1.2	ns

Note: ns, not significant; *, significant at $P < 0.05$

Table 5 Results of RDA analyses of soil chemical properties estimates performed separately for each year. % explanatory variable, concentrations, variability explained by one (all) ordination axis (measures of explanatory power of the explanatory variables); F-ratio, F-statistics for the test of analysis; P-value, probability value obtained by the Monte Carlo permutation test. Tested hypothesis: there is any effect of treatment on soil chemical properties for each year. Applied treatments were described in Fig. 3

Year	Explanatory variables	% explanatory var. 1st axis (all axes)	F-ratio 1st axis (all axes)	P-value 1st axis (all axes)
2013	UM, VA, CO, EG, FP	11.8 (23.9)	2.4 (1.1)	<0.001 (<0.001)
2014	UM, VA, CO, EG, FP	34.3 (30.7)	8.7 (3.9)	<0.001 (<0.001)
2015	UM, VA, CO, EG, FP	23.6 (21.3)	12.5 (3.6)	<0.31 (0.045)
2016	UM, VA, CO, EG, FP	45.2 (47.8)	11.2 (3.8)	<0.001 (<0.001)

Discussion

Soil Chemical Properties

The chemical elements contained in the treatment materials and the prevailing climate substantially influenced the soil chemical properties and consequently exerted significant effects on the different treatments. For example, the

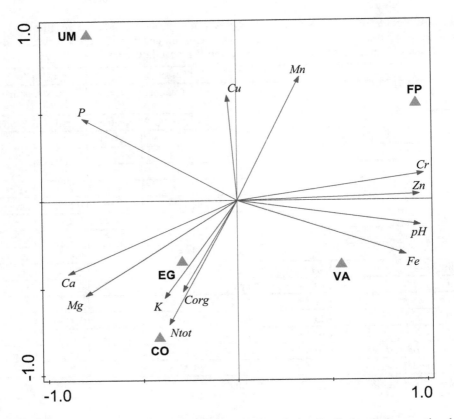

Fig. 5 Ordination diagram showing result of the RDA analysis of soil chemical properties data collected in 4 years (from 2013 to 2016) in different treatments. Treatment abbreviations are explained in Fig. 3

application of *Vernonia amygdalina* and fipronil (FP) elevated the concentrations of most trace elements in the soil, while *Chromolaena odorata* increased the contents of soil organic carbon and other minerals especially in the favorable climate years. Several studies have reported the effects of the exotic plants and organic residues of *V. amygdalina*, *C. odorata*, and *Elaeis guineensis* on the soil chemical properties and microbes (Kushwaha et al. 1981; Obatolu and Agboola 1993; Quansah et al. 2001; Peveling et al. 2003; Koutika et al. 2004; Callaway et al. 2004; Gbaruko and Friday 2007; Banful and Hauser 2011; Tondoh et al. 2013; Agbede et al. 2014; Gandahi and Hanafi 2014; Nurulita et al. 2014; Ngo-Mbogba et al. 2015; Ajayi et al. 2016; Dawson and Schrama 2016; Veldhuis et al. 2017).

It was discovered in the study that VA and FP treatments recorded relatively high soil pH, and this might be attributed to the high concentrations of the trace elements. This finding was consistent with other studies which have revealed that the elevation of heavy metals in most agricultural soils often leads to soil alkalinity (Nederlof et al. 1993; Lenart and Wolny-Koładka 2013; Almaroai et al. 2014) and V. *amygdalina* like *Baphia nitida* accumulates high heavy metals in the region (Ogbonna et al. 2013). On

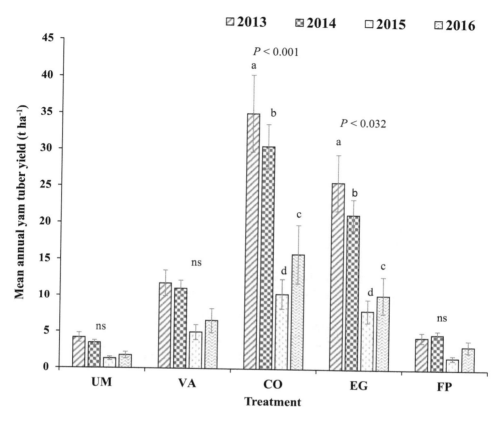

Fig. 6 Mean annual yam tuber yield (t ha^{-1}). Error bars represent standard error of the mean (SEM). *P*-value represents corresponding probability value. *ns* indicates that the results of ANOVA analyses were not significant. Significance differences ($P < 0.05$) between treatments in accordance with the Tukey's post hoc test are shown by lowercase letters (a > b > c > d). Treatment abbreviations were explained in Fig. 3

the contrary, other authors have reported high concentrations of exchangeable Cd and Zn in agricultural soils because of decrease in pH (Sumi et al. 2014).

Although *C. odorata* has been established to be a threat to the soil minerals by some authors (Muniappan et al. 2005), in contrast, several studies in West Africa have positive records about *C. odorata* (Obatolu and Agboola 1993; Goyal et al. 1999; Quansah et al. 2001; Tondoh et al. 2013; Agbede et al. 2014). In agreement with our study, CO treatment revealed increased concentrations of Corg, Ntot, K, and Ca especially in the optimal precipitation years. This might be explained by the deep rooting system of *C. odorata* that mobilizes soil mineral nutrients which are turned into organic and plant-available nutrients in conducive climate (Kushwaha et al. 1981; Tondoh et al. 2013). Other possible reasons for high contents of minerals in the *C. odorata* soil could be due to high contents of leaf biomass and earthworms (Tian et al. 2000; Kone et al. 2012), fast decomposing rate of *C. odorata* (Roder et al. 1995), and the elevated activities of the soil microbes (Mboukou-Kimbasta et al. 2007; Banful and Hauser 2011; Ngo-Mbogba et al. 2015; Dawson and Schrama 2016).

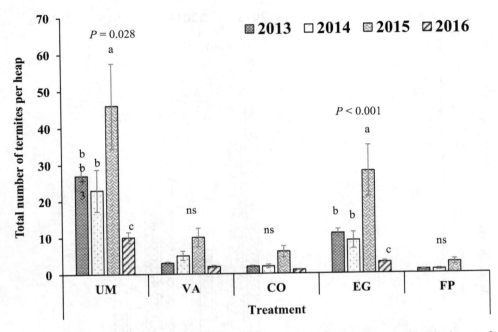

Fig. 7 Total number of termites (*Isoptera*) per yam soil heap. Error bars represent standard error of the mean (SEM). *P*-value represents corresponding probability value. *ns* indicates that the results of ANOVA analyses were not significant. Significance differences ($P < 0.05$) between treatments in accordance with the Tukey's post hoc test are shown by lowercase letters (a > b > c > d). Treatment abbreviations were explained in Fig. 3

Table 6 Relationships between yam tuber yield (kg heap^{-1}) and number of termites per yam heap for the years under the different treatments

Treatment	Equation	R^2	P-value
UM	Y = − 0.0059X + 17.593	0.50	0.045
VA	Y = − 0.014X + 18.824	0.21	0.079
CO	Y = − 0.2456X + 35.04	−0.87	<0.001
EG	Y = − 0.064X + 29.618	−0.75	0.012
FP	Y = − 0.0128X + 15.53	0.24	0.063

On the other hand, *E. guineensis* (EG) treatment had elevated soil organic carbon and total nitrogen in our study. This result was consistent with many studies which reported that the oil palm residues promote the scavenging activities of ants which enrich the soil minerals (Frouz et al. 1997; Gandahi and Hanafi 2014; Nurulita et al. 2014; Gray et al. 2015).

No significant effect of treatment on the soil chemical properties was recorded in 2015 as shown by the RDA analysis. The low amount of rainfall recorded in 2015 contributed to the insignificant role of treatments on the soil because the elements from the treatments required enough soil moisture to show reasonable influence on the soil chemistry (Fernelius et al. 2017).

Yam Tuber Yield and Termites

The highest yam tuber yields were found in the CO treatment. This might probably be explained by high Corg, Ntot, P, K, Ca, and Mg in the soil under the CO treatment which consequently increased the crop yield (Obatolu and Agboola 1993; Quansah et al. 2001; Tondoh et al. 2013; Agbede et al. 2014). Similarly, 2013 and 2014 had the highest yam tuber yields across the different treatments. This was attributed to the optimum rainfall in 2013 and 2014 which did not only favor the activities of the microbial organisms but also helped to decrease the number of the termites' attack on the yam tubers. Though the oil palm (EG) liquid sludge attracted more termites in the soil yet, the EG treatment had higher yam tuber yields than UM and VA treatments. This could be because the termites fed more on the oil palm sludge instead of the yam tubers. Besides, the termites' activities improved the soil minerals (Oviasogie and Aghimien 2003; Muhrizal et al. 2006; Guo et al. 2007; Wu et al. 2009).

The total number of termites per yam heap were significant under UM and EG treatments. This was because of the suitability of the unmanaged and *E. guineensis* (EG) treatments which promoted the termites' activities when compared with the VA, CO, and FP treatments that either constrained or killed the ants (Peveling et al. 2003). Though without high yam yields, fipronil application was the easiest method to eradicate the termites from either destroying the yam tubers or carrying out any activities in the yam farmland. But the risk of fipronil on the soil, environment, and human health has been reported by many authors (Keefer and Gold 2014; Sánchez-Bayo 2014; Lopez-Antia et al. 2016).

Limitations of the Study

Some of the limitations for the study are:

(i) Poor access roads and footpaths to the farm sites where the studies are conducted
(ii) Destruction of some established experiments by mammals such as rabbits (*Oryctolagus cuniculus*), rats (*Rattus*), grasscutters (*Thryonomys swinderianus*), antelopes (*Bovidae*), and others
(iii) Ability to convince the rural farmers to give their farmlands for such research because some of the farmers are afraid that their land might be taken by the government or the research institutes
(iv) Inadequate fund to perform intensive studies: labor, field, laboratory and statistical analyses, and publications
(v) Dearth of literature on the topic in Nigeria and lack of robust data on the climate and other variables

Conclusion

Climate change with its variability was found as a factor which has important influence on yam tuber production because (i) the termite colonies increased when there is low rainfall, (ii) the damage caused by the pests exacerbates during dry season than in wet season, and (iii) optimal rainfall enhances soil moisture and promotes the soil microbial activities and decomposition of organic matter which in turn elevates the soil fertility. *Chromolaena odorata* (CO) treatment has been found to be the best single option in managing the termites' infested farmland to increase yam yield. On the other hand, integrating CO and EG treatment is recommended as this might produce higher yield. Though the termites (pests) were successfully reduced under the VA and FP treatments, this reduction was not commensurate with the yam tuber yield. The high content of the trace elements in the VA and FP treatments was attributed to the reason for declined yields because the fipronil is a good soil-binding substance which limits organic decomposition and nutrient cycling. It is also important to state here that a 4-year study is a short term to conclude that *C. odorata* and *E. guineensis* sludge are sustainable for reducing the yam pests (termites) and their harmful activities.

Since climate change is expected to continue locally and globally, this will have further influence on the distribution and intensification of termites which in turn affects agricultural resources. There is crucial need for the effectiveness of termite management strategies which include integrated pest control methods. Furthermore, as most of the termites are soil inhabitants, soil degradation and low rainfall are the main factors promoting their distribution and forage behavior in both tropical and subtropical ecosystems. Thus, there are high chances for aggravated economic devastation of termites' outbreaks in the region due to climate change. Therefore, the experiment needs to be extended to other locations in the study region. It also requires an intensive and long-term investigation in order to thoroughly understand (i) the influence of climate change on the termites' outbreak, (ii) the extent of termite damage to the crops, (iii) the impacts of climate change and variability on yam yields, (iii) the agricultural and economic benefits of the applied treatments, and (iv) the ecological and human health safety of the treatments.

Acknowledgements The support from the Department of Forest Protection and Entomology, Faculty of Forestry and Wood Sciences, Czech University of Life Sciences, Czech Republic, and the grants "EVA4.0" No. CZ.02.1.01/0.0/0.0/16_019/0000803 from the EVA 4.0 project are appreciated.

Appendix

See Figs. 8, 9, 10, 11, and 12.

Fig. 8 Specimen of *Vernonia amygdalina* (VA) used for the experiment

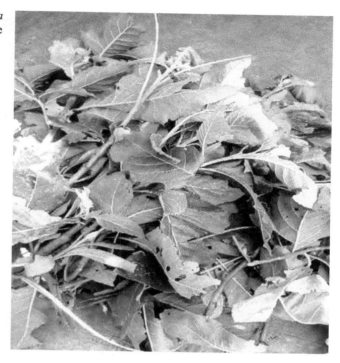

Fig. 9 Specimen of *Chromolaena odorata* (CO) used for the experiment

Fig. 10 Specimen of *Elaeis guineensis* (EG) and its liquid sludge used for the experiment

Fig. 11 Processing of the liquid/sludge of *Vernonia amygdalina* (VA), *Chromolaena odorata* (CO), *Elaeis guineensis* (EG)

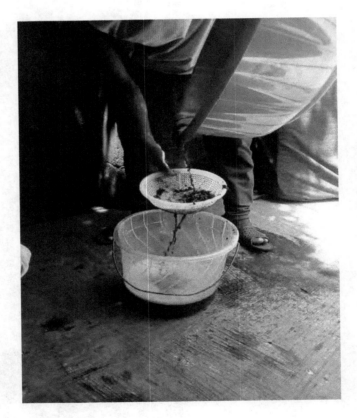

Fig. 12 Stored liquid/sludge of *Vernonia amygdalina*, *Chromolaena odorata*, *Elaeis guineensis* before they were applied in the experiment

References

Agbede TM, Adekiya AO, Ogeh JS (2014) Response of soil properties and yam yield to *Chromolaena odorata* (Asteraceae) and *Tithonia diversifolia* (Asteraceae) mulches. Arch Agron soil sci 60:209–224

Ahmed BM, French JRJ (2008) An overview of termite control methods in Australia and their link to aspects of termite biology and ecology. Pak. Entomol 30:1–18

Ahmed MB, Nkunika POY, Sileshi GW, French JRJ, Nyeko P, Jain S (2011) Potential impact of climate change on termite distribution in Africa. Br J Environ Climate Chang 1(4):172–189

Ajayi FT, Omotoso SO, Odejide JO (2016) Evaluation of fodder plants (*Ficus polita*, *Azadirachta indica* and *Vernonia amygdalina*) for their phytochemical and antibacterial properties. Cog. Food Agric 2(1):1211466

Almaroai YA, Vithanage M, Rajapaksha AU et al (2014) Natural and synthesized iron-rich amendments for As and Pb immobilization in agricultural soil. Chem Ecol 30(3):267–279

Asante SK, Mensah GWK, Wahaga E (2008) Farmers' knowledge and perceptions of insect pests of yam (*Dioscorea* spp.) and their indigenous control practices in northern Ghana. Ghana J Agric Sci 40(2):185–192

Atu UG (1993) Cultural practices for the control of termite (*Isoptera*) damage to yams and cassava in south-eastern Nigeria. Intern J Pest Manag 39:462–466

Banful B, Hauser S (2011) Changes in soil properties and nematode population status under planted and natural fallows in land use systems of southern Cameroon. Agrofor Syst 82:263–273

Beaudrot L, Du Y, Rahman Kassim A, Rejmanek M, Harrison RD (2011) Do epigeal termite mounds increase the diversity of plant habitats in a tropical rain forest in peninsular Malaysia? PLoS ONE 6(5):e19777

Belyaeva NV, Tiunov AV (2010) Termites (*Isoptera*) in forest ecosystems of Cat Tien national park (Southern Vietnam). Biol Bullet 37:374–381

Bignell DE, Eggleton P (2000) Termites in ecosystems. In: Abe T, Bignell DE, Higashi M (eds) Termites: evolution, sociality, symbioses, ecology. Kluwer Academic Publications, Dordrecht, pp 363–387

Boko MI, Nyong NA, Vogel C, Githeko A, Medany M, Osman-Elasha B, Tabo R, Yanda P (2007) Africa climate change 2007: impacts, adaptation and vulnerability. Contribution of Working Group II to the fourth assessment report of the Intergovernmental Panel on Climate Change, Parry ML, Canziani OF, Palutikof JP, van der Linden PJ, Hanson CE (eds). Cambridge University Press, Cambridge UK, pp 433–467

Boonyatumanond R, Tabucanon MS, Thongklieng S, Boonchlaermkit S (2006) Persistent of organochlorine pesticides in the green Lip Mussel (*Perna Viridis*) from marine Estuaries in Thailand. Chem Ecol 17(1):31–39

Bremner JS, Mulvaney CS (1982) Nitrogen-total. In: Page AL (ed) Methods of soil analysis no. 9, Part 2, Chemical and microbiological properties. American Society of Agronomy, Madison

Buczkowski G, Bertelsmeier C (2017) Invasive termites in a changing climate: a global perspective. Ecol Evol 7:974–985

Callaway RM, Thelen GC, Rodriguez A, Holben WE (2004) Soil biota and exotic plant invasion. Nature 427:731–733

Christensen JH, Hewitson B, Busuioc A, Chen A, Gao X, Held I, Jones R, et al (2007) Regional climate projections. Climate change 2007: the physical science basis. Contribution of Working Group I to the Fourth Assessment Report of the Intergovernmental Panel on Climate Change, Solomon S, Qin D, Manning M, Chen Z, Marquis M, Averyt KB, Tignor M, Miller HL, Eds., Cambridge University Press, Cambridge, pp 847–940

Clayton PM, Tiller KG (1979) A chemical method for the determination of heavy metal content of soils in environmental studies: CSIRO, Australia Melbourne Division. Soil Technol 41:1–7

Consultative Group on International Agricultural Research (CGIAR) (2004) Research and impact: area of research: yam. C:\User\user\Documents\Yam-CGIAR Research & Impact Areas of Research Yam.mh

Dawson W, Schrama M (2016) Identifying the role of soil microbes in plant invasions. J Ecol 104:1211–1218

Devendra G, Ayo-Odongo J, Kit V, et al (1998) A participatory systems analysis of the termite situation in West Wollega, Oromiya region, Ethiopia. Working document series 68: ICRA/EARO/OADB (Oromia Agricultural Development Bureau), p 168

Doust H, Omatsola E (1990) Niger delta. In: Edwards JD, Santogrossi I (eds) Divergent and passive margin basins. AAPG Memoir, vol 48: Tulsa, American Association of Petroleum Geologists, Tulsa, pp 201–238

Eggleton P, Bignell DE, Hauser S, Dibog L, Norgrove L, Madong B (2002) Termite diversity across an anthropological gradient in the humid forest zone of West Africa. Agric Ecosyst Environ 90:189–202

Fernelius KJ, Madsena MD, Hopkins GB et al (2017) Post-fire interactions between soil water repellency, soil fertility and plant growth in soil collected from a burned piñon-juniper woodland. J Arid Environ 144:98–109

Frouz J, Santruckova H, Kalcik J (1997) The effect of wood ants (Formica polyctena Foerst.) on the transformation of phosphorus in a spruce plantation. Pedobiologia 41:437–447

Gandahi AW, Hanafi MM (2014) Bio-composting oil palm waste for improvement of soil fertility. Part of the Sustainable development and biodiversity book series (SDEB). Composting Sustain Agric 3:209–243

Gbaruko BC, Friday OV (2007) Bioaccumulation of heavy metals in some fauna and flora. Intern J Environ Sci Technol 4:197–202

Goyal S, Chander K, Mundra MC, Kappor KK (1999) Influence of inorganic fertilizers and organic amendments on soil organic matter and soil microbial properties under tropical conditions. Biol Fert Soil 29:196–200

Gray CL, Lewis OT, Chung AYC et al (2015) Riparian reserves within oil palm plantations conserve logged forest leaf litter ant communities and maintain associated scavenging rates. J Appl Ecol 52:31–40

Guo TR, Zhang GP, Zhang YH (2007) Physiological changes in barley plants under combined toxicity of aluminum, copper and cadmium. Colloids Surf B Biointerfaces 57:182–188

Harrington R, Stork NE (eds) (1995) Insects in a changing environment. Academic, London

IBM Corporation (2011) IBM SPSS statistics for windows, version 20.0. Armonk: IBM Corporation. www.ibm-spss-statistics.soft32.com

Jones JA (1990) Termites, soil fertility and carbon cycling in dry tropical Africa: a hypothesis. J Trop Ecol 6:291–305

Keefer CT, Gold RE (2014) Recovery from leachate and soil samples of fipronil at termiticide concentration. Southwestern Entomol 39:705–716

Kemp PB (1955) The termites of North-Eastern Tanaganyika (Tanzania): their distribution and biology. Bull Entomol Res 46:113–135

Kone AW, Edoukou EF, Gonnety JT et al (2012) Can the shrub *Chromolaena odorata* (Asteraceae) be considered as improving soil biology and plant nutrient availability? Agrofor Syst 85:233–245

Koutika LS, Hauser S, Meuteum K et al (2004) Comparative study of soil properties under *Chromolaena odorata*, *Pueraria phaseoloides* and *Calliandra calothyrsus*. Plant Soil 266:315–323

Kurukulasuriya P, Mendelsohn R, Hassan R, Benhin J, Deressa T, Diop M, Dinar A et al (2006) Will African agriculture survive climate change? World Bank Econ Rev 20:367–388

Kushwaha SPS, Ramakrishnan PS, Tripathi RS (1981) Population dynamics of *Eupatorium odoratum* in successional environments following slash and burn agriculture. J Appl Ecol 18:529–535

Lenart A, Wolny-Kołádka K (2013) The effect of heavy metal concentration and soil pH on the abundance of selected microbial groups within ArcelorMittal Poland Steelworks in Cracow. Bull Environ Contam Toxicol 90:85–90

Logan JWM, Cowie RW, Wood TG (1990) Termite (Isoptera) control in agriculture and forestry by nonchemical methods: a review. Bull Entomol Res 80:309–330

Loko YL, Agre P, Orobiyi A et al (2015) Farmers' knowledge and perceptions of termites as pests of yam (*Dioscorea spp.*) in Central Benin. Intern J Pest Manag 62(1):75–84

Lopez-Antia A, Feliu J, Camarero PR et al (2016) Risk assessment of pesticide seed treatment for farmland birds using refined field data. J Appl Ecol 53:1373–1381

Lovett JC, Midgley GF, Barnard P (2005) Climate change in Africa. Afr J Ecol 43:167–169

Mboukou-Kimbasta I, Bernhard-Reversat F, Loumeto JN et al (2007) Understorey vegetation, soil structure and soil invertebrates in Congolensis Eucalypt plantations, with special reference to the invasive plant *Chromolaena odorata* and earthworm populations. Eur J Soil Biol 47:48–56

McLean EO (1982) Soil pH and lime requirements. In: Page AL (ed) Methods of soil analysis. Part 2. Chemical and microbiological properties, 2nd edn. Agronomy series, 9. ASA, SSSA, Madison

Mehlich A (1984) Mehlich III soil test extractant: a modification of Mehlich 2 extractant. Commun Soil Sci Plant Anal 15:1409–1416

Muhrizal S, Shamshuddin J, Fauziah I et al (2006) Changes in iron-poor acid sulphate soil upon submergence. Geoderm 131:110–122

Muniappan R, Reddy GVP, Lai PY (2005) Distribution and biological control of *Chromolaena odorata*. In: Inderjit S (ed) Invasive plants: ecological and agricultural aspects. Birkhauser Verlag AG, Basel, pp 223–233

Nederlof MM, Van Riemsdijk WH, De Haan FAM (1993) Effect of pH on the bioavailability of metals in soils. In: Eijsackers HJP, Hamers T (eds) Integrated soil and sediment research: a basis for proper protection. Springer, Kluwer Academic Publishers, pp 215–219

Nelson DW, Sommers LE (1996) Total carbon, organic carbon and organic matter. In: Sparks DL (ed) Methods of soil analysis. Part 3, 2nd edn. SSSA book series no. 5. ASA and SSSA, Madison, pp 961–1010

Ngo-Mbogba M, Yemefack M, Nyeck B (2015) Assessing soil quality under different land cover types within shifting agriculture in South Cameroon. Soil Tillage Res 150:124–131

Nkunika POY (1998) Termite survey, identification, damage and control in southern province. Land Management and Conservation Farming, Soil Conservation and Agro forestry Extension Report, Lusaka, Zambia

Nurulita Y, Adetutu E, Kadali K et al (2014) The assessment of the impact of oil palm and rubber plantations on the biotic and abiotic properties of tropical peat swamp soil in Indonesia. Int J Agric Sustain 13(2):150–166

Nwankwoala HO, Nwaogu C (2009) Utilizing the tool of GIS in oil spill management-a case study of Etche LGA, Rivers State, Nigeria. Glob J Environ Sci 8:19–29

Nwaogu C, Ogbuagu HD, Abrakasa S et al (2017) Assessment of the impacts of municipal solid waste dumps on soils and plants. Chem Ecol 33(7):589–606

Obatolu CR, Agboola AA (1993) The potential of Siam weed (*Chromolaena odorata*) as a source of organic matter for soils in humid tropics. In: Mulongoy M, Merckx R (eds) Soil organic matter dynamics and sustainability of tropical agriculture. Wiley-Sayce Co, New York, pp 89–99

Ogbonna PC, Odukaesieme C, Teixeira da Silva JA (2013) Distribution of heavy metals in soil and accumulation in plants at an agricultural area of Umudike, Nigeria. Chem Ecol 29(7):595–603

Olajumoke OL, Agiang MA, Mbeh E (2012) Proximate and anti-nutrient composition of white Guinea yam (*Dioscorea rotundata*) diets consumed in Ibarapa, South West region of Nigeria. J Nat Prod Plant Resour 2:256–260

Olorede KO, Alabi MA (2013) Economic analysis and modelling of effects of NPK fertilizer levels on yield of yam. Math Theory Model 3:108–118

Osunde ZD, Orhevba BA (2009) Effects of storage conditions and storage period on nutritional and other qualities of stored yam (*Dioscorea spp.*) tubers. African Journal of Food, Agriculture. Nutr Dev 9:678–690

Oviasogie PO, Aghimien AE (2003) Macronutrient status and speciation of Cu, Fe, Zn and Pb in soil containing palm oil mill effluent. Glob J Pure Appl Sci 9:71–80

Peveling R, McWilliam AN, Nagel P et al (2003) Impact of locust control on harvester termites and endemic vertebrate predators in Madagascar. J Appl Ecol 40:729–741

Pomeroy DE (1976) Studies of population of large termite mounds in Uganda. Ecol Entomol 1:49–61

Quansah C, Fening JO, Ampontuah EO et al (2001) Potential of *Chromolaena odorata, Panicum maximum* and *Pueraria phaseoloides* as nutrient sources and organic matter amendments for soil fertility maintenance in Ghana. Biol Agric Hortic 19:101–113

Richard JF, Ait-Baddi G, Costa C et al (2006) Comparative study of humic acids of the mound of a wood-feeding termite and of the litter directly below in the Amazon river delta. Chem Ecol 22(3):201–209

Roder W, Phengchanh S, Keoboualapha B et al (1995) *Chromolaena odorata* in slash-and-burn rice systems of Northern Laos. Agrofor Syst 31:79–92

Rouland-Lefèvre C (2011) Termites as pests of agriculture. In: Bignell DE et al (eds) Biology of termites: a modern synthesis. Springer Science + Business Media B.V, Dordrecht, pp 499–517

Sánchez-Bayo F (2014) The trouble with neonicotinoids: chronic exposure to widely used insec-ticides kills bees and many other invertebrates. Science 346(6211):806–807

Sartie A, Franco J, Asiedu R (2012) Phenotypic analysis of tuber yield- and maturity-related traits in white yam (*Dioscorea rotundata*). Afr J Biotechnol 1:3964–3975

Sekamatte MB, Okwakol MJN (2007) The present knowledge on soil pests and pathogens in Uganda. Afri J Ecol 45(2):9–19

Shell Petroleum Development Company of Nigeria Limited (SPDC) (1998) Environmental impact assessment of Obigbo Node associated gas gathering project: final report by Tial Trade Limited, Nigeria

Sileshi GW, Nyeko P, Nkunika POY, Sekamatte BM, Akinnifesi FK, Ajayi OC (2009) Integrating ethno-ecological and scientific knowledge of termites for sustainable termite management and human welfare in Africa. Ecol Soc 14(1):48

Slingo JM, Challinor AJ, Hoskins BJ, Wheeler TR (2005) Food crops in a changing climate. Philos Trans R Soc B Biol Sci 360:1983–1989

Šmilauer P, Lepš J (2014) Multivariate analysis of ecological data using CANOCO 5, 2nd edn. Cambridge University Press, Cambridge

Sumi H, Kunito T, Ishikawa Y et al (2014) Plant roots influence microbial activities as well as cadmium and zinc fractions in metal-contaminated soil. Chem Ecol 31(2):105–110

Tian G, Kolawole OG, Kang BT et al (2000) Nitrogen fertilizer replacement indexes of legume cover crops in the derived savannah of West Africa. Plant Soil 224:287–296

Tondoh JE, Kone AW, N'Dri JK et al (2013) Changes in soil quality after subsequent establishment of *Chromolaena odorata* fallows in humid savannahs, Ivory Coast. Catena 101:99–107

UNEP (2000) FAO and Global IPM facility expert group on termite biology and management. United Nations Environmental Programme: chemicals finding alternatives to Persistent Organic Pollutants (POPs) for termite management. http://www.chem.unep.ch/Publi cations/pdf/Alterna tives-termitefulldocument.pdf

Veldhuis MP, Laso FJ, Olff H et al (2017) Termites promote resistance of decomposition to spatiotemporal variability in rainfall. Ecology 98:467–477

Wood TG (1995) The agricultural importance of termites in the tropics. Agric Zool Rev 7:117–155

Wu TY, Mohammad AW, Jahim JM et al (2009) A holistic approach to managing palm oil effluent (POME): Biotechnological advances in the sustainable reuse of POME. Biotechnol Adv 27:40–52

Zimmerman PR, Greenberg JP (1983) Termites and methane. Nature 302:354–355

9

Unlocking Climate Finance Potential for Climate Adaptation: Case of Climate Smart Agricultural Financing in Sub-Saharan Africa

Edward M. Mungai, S. Wagura Ndiritu and Izael da Silva

Contents

E. M. Mungai · S. W. Ndiritu
Strathmore University Business School, Nairobi, Kenya
e-mail: emungai@kenyacic.org; sndiritu@strathmore.edu

I. da Silva (✉)
Strathmore University, Nairobi, Kenya
e-mail: idasilva@strathmore.edu

Abstract

Climate change has emerged as one of the greatest challenges faced by the world today. Adverse impacts of climate change are visible across sectors like agriculture and other natural resources due to increasing average temperature and changing weather patterns. Africa constitutes around 13% of the global population but contributes the least (around 2%) to greenhouse gas (GHG) emissions globally. Concerning the global climate vulnerability index, Africa is most impacted (around 21%) by climate change and its' population is most vulnerable to climate sensitivity and fragility of the continent's natural environment and increasingly erratic weather patterns, low adoption of climate-resilient technologies, and high dependence on environment-based livelihoods. Hence, Africa needs to adopt low carbon and climate-resilient development to address climate-related issues and to have sustainable development. In line with the low carbon/climate-resilient development agenda, 53 countries (except Libya) have submitted Nationally Determined Contribution (NDC) and have set ambitious targets under NDC and Sustainable Development Goals. A quick analysis of the NDCs and various studies indicates the enormity of the financing needs. According to Climate Invetsment Funds (CFI), Sub-Saharan Africa will require an estimated USD222 billion for climate resilience investments to reach its NDCs. One of the critical stakeholders to play a key role in meeting the financing needs of climate-smart agriculture (CSA) related targets is the private sector. There is around 98% gap in financing for CSA. Even though substantial climate finance potential exists in selected countries for the private sector, there are certain challenges and barriers like financial, policy, lack of awareness, and low provision for climate funding in the national budget.

Keywords

Climate-smart agriculture · Unlocking climate finance · Sub-Saharan Africa · Climate adaptation · Private sector

Introduction

An additional 2.4 billion people – representing a one-third increase in the global population – will occur between 2013 and 2050 (FAO 2013). Further, the Food and Agriculture Organization (FAO) approximates the additional population will translate into a 60% increase in demand for agricultural production. Undoubtedly, agriculture is well-positioned to be a reliable basis for economic growth and poverty reduction. Conversely, the ongoing global environmental concern of climate change has adverse impacts on agriculture that is also a contributing factor to the drastically changing weather and climatic patterns. Consequently, for agriculture to satisfactorily feed the growing population, the current agriculture practices need to spiral into more adaptive hence sustainable practices. A breakthrough hinged on climate-smart agriculture (CSA) approach that encompasses three perspectives: increasing productivity in a sustainable manner, enhancing adaptation/resilience, and mitigating the emission of greenhouse gases (GHGs) as emerged as a way to realize food security and achieve the developmental goals. Notably, 56% of Africa's population will reside in urban areas by 2025 and half of the projected population growth will be in Sub-Saharan Africa (UNDESA 2019 and 2014). This entails that agriculture, specifically in Africa, has to undergo a major transformation to fulfill the intertwined challenges of achieving food security, reducing poverty, and responding to climate change without depletion of the natural resource base. Despite the common consensus on the potential for climate-smart agriculture in Africa, there is a conspicuous paucity in wholesome quantifiable empirical evidence. To fill this gap, this chapter looks into the potential for climate-smart agriculture in 14 select countries in Sub-Saharan Africa (SSA) representing East, West, Central, and Southern Africa.

World Bank (2015) estimates that about 48% (approximately 450 million people) of Africa's population live in extreme poverty, i.e., less than US $1.25 in a day. Also, 63% of Africa's population lives in rural areas, wholly dependent on agriculture as a source of living (World Bank 2015). More than 60% of the African population works in the agricultural sector that accounts for about 25–34% of the continent's gross domestic production (GDP). Conceivably, agriculture looms large in the African economy. Unfortunately, a collective report by FAO, IFAD, and WFP (2014) revealed that agricultural production is low leading to high food insecurity. A reason attributed to sluggish income growth, high poverty rates, and dilapidated infrastructure in the rural areas that impair market access. A situation further exacerbated by weak policies, civil unrest, periodical disease outbreaks, overlapping rules, poor coordination, and inept collaboration among institutions within the climate-smart agriculture realm. One in four people remains malnourished in Africa with a high prevalence of stunted and underweight children due to poor dietary quality and diversity, mostly among the poor. Increasing agriculture's adaptive capacity will be necessary to prevent a slide back into poverty and hunger.

Growth in agriculture is the most viable and equitable strategy to spur economic growth in Africa by reducing poverty and enhancing food security in Africa. However, it has to overcome the climate change-related challenges. For instance, Barkhordarian et al. (2012), Radhouane (2013), and IPCC (2014) postulated that the annual rainfall in Sub-Saharan Africa will possibly decrease by about 4–47% resulting in droughts and increased salinity. It resonates with Intergovernmental Panel for Climate Change observations that crop and fodder growing periods in both western and southern Africa will likely shorten by an average of 20% by the year 2050. Resultantly, there will be a 40% decline in cereal yield and an additional reduction in cereal biomass for livestock (Lobell et al. 2011). According to Hoerling et al. (2006) western, central, and southern Africa will record a decline in the mean annual rainfall of 4%, 5%, and 5%, respectively. In the rest of Africa, drought conditions will not only be more frequent and intense but also more long-lasting leading to an increase in the arid and semiarid area approximately to about 5–8% by 2080 (Elrafy 2009). As a factor attributed to the sensitivity of the current farming systems to drought, the cumulative crop yield decline across the continent is forecasted at 50% by 2020. Thornton et al. (2008) contend that the net revenues from crops may likely fall by about 90% by 2100. Further, both agropastoral, pastoral, and mixed-crop livestock systems will potentially be affected by a constraint of animal feed and water in addition to advancing pest severity and disease distribution (Thornton et al. 2008). Against such a grim picture, there lies an excellent opportunity with CSA for transformation by collating agriculture, economic growth, and climate change under the umbrella of sustainable development. Four agroecological zones in SSA serves as case studies for underlying CSA investment potential.

To facilitate CSA adoption in developing countries, respective governments' have claimed their right to public grants with lesser regard to private financing (Pauw 2014). The latter predisposition of the developing countries is in line with the United Nations Framework Convention on Climate change (UFCC) principle of "the polluter pays." The approach implies that developed countries should play a greater role in providing climate adaptation finance for being major contributors of climate change. Bindingly, developed countries pledged US $100 billion every year from 2020 onwards (UFCC 2011). A target that Pauw et al. (2015) doubts if it will be met, actually UNEP FI (2009) postulates that public funding cannot sufficiently finance climate change adaptation costs. Pauw (2014) proposes that the private sector can supplement but not substitute public investment in climate finance. Observations by UFCC (2007) approximated the global private sector investment and financial flows at 86%. Further, SER (2011) cite that 90% of the population in emerging economies depends on the private sector as a source of income. Pauw and Pegels (2013) argues that the private sector can play a potentially significant role in adaptation engagement. As a result of the private sector potential, it was included as one of the finance sources. However, as put forward by Surminski (2013), the evidence base – reasonable activity, predictable returns and acceptable risk for private sector investment – is limited (Christiansen et al. 2012).

This chapter will contribute to the extant literature in twofolds. First is a pioneering academic exploration into quantifying the investment potential and the funding gap in climate-smart agriculture in Africa. This is unlike recent work of Tran et al. (2019) that focused on determinants for the adoption of CSA technologies in developing countries and Pauw and Pegels (2013) with a reflection on the role of the private sector in developing countries. This chapter also digresses from a study by Zougmoré et al. (2018) and Nciizah and Wakindiki (2015) that looked into the prospects and the achievements of CSA in Africa. Secondly, unlike past generalizations on the areas that need climate financing, this chapter will identify the financial, regulatory, and policy barriers hindering private sector investment into CSA projects in Sub-Saharan Africa. In general, this work will add to the ongoing research on the conceptual clarity of private sector engagement in climate adaptation in developing countries (Pauw and Pegels 2013; Pauw 2014).

Based on the selected sample, this work found out that the highest climate-smart agriculture finance potential (in USD billion) lay with Ethiopia at USD 26 billion, distantly followed by Nigeria at USD 17 billion, and further down is Kenya at USD 9 billion which is almost the same case for Madagascar at USD 8 billion. Interestingly, CSA was termed as more investor-friendly, receiving a cumulative investment of

USD 79 billion in the 14 countries. Notwithstanding the CSA potential, the sector faces several challenges including inadequate financing, weak policies, and knowledge gaps within the key institutions. Nonetheless, there is a 98% untapped climate-smart agriculture investment potential that the private sector can exploit through climate financing.

The remainder of this chapter proceeds as follows. The second section will provide an elaborate literature review while the following section will look into the methodology that was used to come up with the conclusions and recommendation. Finally, the chapter will conclude by presenting the conclusion and recommendations.

The Concept of Climate-Smart Agriculture

The approach to developing the technical capacity, accommodative policies and create an enabling environment for investment towards sustainable agriculture in the face of climate change was put forward by FAO (2013). CSA simultaneously addresses the global concerns of food security, ecosystem management, and climate change, therefore incorporating the three dimensions of sustainable development: economic, social, and environmental conditions. Nciizah and Wakindiki (2015) identifies three pillars of CSA as: (1) sustainably increasing agricultural productivity from crops, livestock, and fisheries without detrimental effects to the ecosystem, (2) reducing short-term farming shocks while enhancing their resilience by increasing farmers adaptability to long-term stresses, and (3) mitigating (GHG) emissions by either removing or reducing possible pollution instances. At the grassroots, CSA is intended to bolden the livelihoods through food security mostly among the small-scale farmers. This is by improving the management of natural resources and shifting to suitable technological approaches for the production, processing, and marketing farm produce. Relatedly, at the national level, CSA is tailored to prompt mainstreaming of policy, technical, and financial mechanisms that facilitate a base for operationalization of climate change adaptation within the agriculture sector. There is a vast array of CSA technologies that can be used singly or in combination in response to various environmental conditions (Teklewold et al. 2017). But mainly, the adoption of CSA technologies among farmers differs based on cultures, preferences, awareness, socioeconomic backgrounds, and resource availability (Maguza-Tembo et al. 2017). However, successful CSA requires an appropriate match between agricultural production technologies with social, economic, and environmental conditions.

The Potential for Climate-Smart Agriculture in Sub-Saharan Africa

In West Africa, there is a high and fast-growing population and increasing agricultural production intensification to meet the growing food demand is very limited. Buah et al. (2017), Jalloh et al. (2012), and Sanou et al. (2016) postulate that to enhance food security in West Africa there will be a need to have animals that have resilient genetic potential, drought-resistant crop varieties that are also hardly affected by insects and diseases. Besides, the management of soil carbon and fertility techniques will be handy for the region. Jalloh et al. (2012) also note that increasing capabilities among the smallholders and large-scale irrigated farms are likely to open up new opportunities for CSA through approaches such as crop-livestock interactions. CSA opportunities in Central Africa lie from an increasing but also a food-insecure population. In this case, sustainably increasing agricultural productivity will not only enhance food security but will also prevent deforestation. CSA aims at limiting the expansion of cultivated land into forests by seeking alternatives in better agricultural techniques that are productive hence restoring lost ecosystems.

Torquebiau (2015) contends that to enhance food security in East Africa, CSA practices need to put more emphasis on increasing livestock productivity, soil conservation, and management of water and natural resources at both landscape and small-scale levels, adaptive intensification of the cropping systems. Further, Partey et al. (2016) and Zougmoré et al. (2015) observed that CSA innovations in East Africa should expound into agroforestry, development of stress-tolerant crops and livestock, crop-livestock diversification, as well as combining conservation agriculture with integrated soil fertility management. For instance, research by Wambugu et al. (2011) in Western Kenya found out that short-term agroforestry fallows being used in some parts of Western Kenya increased the annual net income of farmers to between US $62 and 122. Such interventions need to be adopted across Africa although with regional-suited species.

According to Mapfumo et al. (2015), South Africa rainfall is expected to decrease and incidences of drought to increase just like other parts of Africa. Similarly, like East Africa, South Africa has to increase its agricultural productivity through intensification. The most crucial CSA approaches in South Africa will be integrated soil, water, nutrient, and organic manure management (Mapfumo et al. 2015). Additionally, soil carbon, salinity, and organic matter regulation will be critical gains for CSA in empowering the smallholder communities to overcome the food shortages and nutrient scarcity. Mbow et al. (2014) also advocated for the use of legume cereal rotational systems in South Africa that should be combined with inorganic fertilizers.

Majority of the Intended Nationally Determined Contributions (INDCs) have referenced agriculture as an adaptation priority, despite that in most cases, there are no cost estimates and adequate financial mechanisms (World Bank 2016). Of critical value in realizing the adaptation objectives is increasing the working and investment capital in climate-smart agriculture. In 2014, the total climate finance mobilized globally was US $391, and despite agriculture's vulnerability to climate change, only US $6–8 billion was committed to livestock, fisheries, and crops. Ironically, the total financing demand for smallholder farmers in Latin America, Asia, and Sub-Saharan Africa was estimated at $210 billion. Consequently, agriculture in developing countries has a challenge of access to sufficient and adequate finance due to high and perceived risks, low margins for financiers, and profitability. As a result, financiers limit their exposure, raise the interest rates, tighten the lending requirements, shorten lending durations, and others opt for other economic sectors with stable returns (World Bank 2016). Among the factors contributing to the funding gap are imbalanced risk-reward profiles, limited capacity to identify financial needs for adaptation, and insufficient evidence bases to identify suitable climate-smart practices and potential. Therefore, there is need to hypothesis that *there is a substantial climate-smart agriculture investment potential in African countries.*

Climate-Smart Agriculture Opportunities and Impeding Challenges in Sub-Saharan Africa

In June 2014, leaders from African Union member states endorsed the adoption of CSA in the New Partnership for Africa's Development (NEPAD). Further, the summit crafted the African Climate-Smart Agriculture Alliance whose aim was to partner with regional economic communities and nongovernmental organizations in enhancing NEPAD planning and coordination to impact on 25 million farm households by the year 2025. Progressively, ECOWAS (2015) and Zougmoré et al. (2015) note that Economic Community of West African States (ECOWAS) initiated the West Africa CSA Alliance to imbed climate-smart agriculture within the programs of ECOWAS Agricultural Policy (ECOWAP)/Comprehensive Africa Agricultural Development Program (CAADP). Respective heads of states that are signatory to NEPAD agreed to a collaboration between NEPAD Planning and Coordinating Agency (NPCA) and the nongovernmental organizations aimed at boosting agricultural productivity by boldening climate change adaptive capacity at the grassroots level. According to the African Union, the ensuing partnership was to avail technical aid to AU members to enhance CSA implementation. Similarly, the African Development Bank (ADB) together with partners were to support African countries on investing in CSA. Using FAO guidelines, several African countries have identified specific agriculture investment needs for the upscaling of CSA implementation (FAO 2012). More so, they have revised their National Agriculture Investment Plans.

On the analysis of the National Adaptation Plans of Action (NAPA) among the 47 least developed countries, about half (22) of the countries explicitly recognized the needed role from private sector (Pauw and Pegels 2013). According to the study, some of the countries broadly identified areas of engagement for the private sector. The countries view the private sector as a partner in the adaptation of sustainable sources of energy – specifically transition from wood and charcoal into solar and wind, agricultural practices, and water management. However, only one country (Mali) recognized the cofinancing role of the private sector (Pauw and Pegels 2013), showing low levels of awareness of the role of the private sector in climate adaptation. Alternatively, the failure to recognize the private sector in the NPAs may be intentional delusion to facilitate the scaling up of public funding. Altogether, 90% of the NPA recognized inadequate finance resources as a potential barrier to climate adaptation, and at the same time, only about 10% presented lack of private sector engagement as a barrier (Pauw and Pegels 2013).

Reportedly, national organizations have rapidly embraced CSA implementation. However, CSA is at its infancy due to a myriad of problems. Specifically, Barnard et al. (2015) present limited access to credit and finance as a major obstacle towards the adoption of CSA practices as they hinder access to farm tools and inputs. Further, Milder et al. (2011) argue initial investment into CSA is prohibitive especially for small farmers that according to Branca et al. (2012) constitute the largest share of agriculture investment in Africa. Mhlanga et al. (2010) reported that investment in agriculture by banks in Africa is barely 10% and attracts relatively high-interest rates. However, to unlock such potential, there is a need to carefully detail the barriers that may hold back CSA development in Africa.

Partey et al. (2016) noted that there is a limited understanding of the CSA concept and framework. All across Africa, farming practices and systems differ creating uncertainties into what technologies or activities constitute CSA. As the advocacy for CSA grows, essential stakeholders such as financial institutions fail to recognize their role in influencing the smart agriculture initiative hence failing to promote the scaling up investment. Williams et al. (2015) note that initially, there were policies, strategies, plans, and programs that were formulated without being informed by the concept of CSA. This has resulted in incompatibility challenges and at times leads to policy duplication. As an observation by Williams et al. (2015), the majority of the West African countries are yet to integrate climate change adaptation into their respective country's national agriculture programs. Additionally, there is limited investment in CSA due to a narrow number of technological packages and financial products (Partey et al. 2016). A factor that can be attributed to limited economic documentation of CSA implications that lead to a failed business case to attract investment (Sylla et al. 2012; Giller et al. 2009). Pauw and Pegels (2013) noted that attracting adaptation investment from the private sector may be challenging in developing countries due to constrained business environment, underdeveloped private sectors and lack of experience with adaptation engagement among the private sector. To help in the risk analysis necessary for private investment decision-making, there is need to hypothesize that there are *major challenges that hinder private sector investment in climate-smart agriculture in Africa.*

Countries Selection

This section describes the process of determining the country of focus for the assessment of the CSA financial potential as well as the barriers to CSA financing in Sub-Saharan Africa. The work carried out was conducted in 14 countries: 4 from East, West, and Southern Africa and an additional 2 countries from Central Africa. The countries are about 30% representative of the four different agroecological regions in Africa. Table 1, depicts the number of countries in each region and the number of those shortlisted.

Table 1 Number of shortlisted countries from each geographical region in Africa

S. No.	Sub-Saharan African region	Total number of countries	Number of countries shortlisted for climate finance study
1	Central Africa	7	2
2	East Africa	13	4
3	Southern Africa	14	4
4	West Africa	15	4

Sampling Selection Method

The shortlisted countries were based on five key indicators related to climate change. First, as the study relates to unlocking private capital, the *foreign direct investment* (FDI) inflows into each country was factored in. Secondly, a consideration was given to the *climate risk score* to enable the inclusion of the most vulnerable countries in detailed research. Thirdly, desktop research was undertaken to identify the *climate finance requirements* for each of the countries. Fourthly, it was a consideration of the *ease of doing business* to provide key insight into government initiatives. Lastly, the *GDP and its growth curve* were put into consideration to determine the demand for each country. In each indicator, countries were ranked from the highest to the lowest as shown in Table 2.

The data for each indicator was collected from credible sources such as the World Bank and the International Finance Corporation. The data on climate risk score was retrieved from the Global Climate Risk Index score released by Germanwatch. Nationally Determined Contributions that were submitted by each country as per the Paris Agreement were reviewed to determine the climate finance requirement. Data from the Germanwatch that monitors the impacts of weather-related loss events were used to develop the country ranks for climate risk index. Climate Risk Index data for 2017 was used as it was the most recent.

Situational Analysis of Priority Countries

After the indicator-based ranking, comprehensive desk research was carried out on climate vulnerabilities, climate change scenarios, national priorities, and policies related to climate change adaptation. Subsequently, a review of Nationally Determined Contributions and Sustainable Development Goals (SDGs) of the selected countries was conducted to identify climate adaptation investment potential.

Estimating Climate Investment Potential

To estimate the investment potential for climate-smart agriculture up to 2030, there was detailed desk research of documents like the National Adaptation Plan Actions, Climate-Smart Agriculture – country factsheets, Economic Cost of Adaptation,

Table 2 Ranking of countries based on the key indicators

Region/parameter	GDP (current USD Billion)	FDI inflow (USD Million)	Climate risk score	Climate finance requirement	Ease of doing business	GDP growth rate (%)
East Africa	1. Sudan 2. Ethiopia 3. Kenya 4. Tanzania	1. Ethiopia 2. Tanzania 3. Sudan 4. Uganda	1. Somalia 2. Kenya 3. Ethiopia 4. Sudan	1. Ethiopia 2. Tanzania 3. Kenya 4. Rwanda	1. Rwanda 2. Kenya 3. Seychelles 4. Uganda	1. Ethiopia 2. Tanzania 3. Rwanda 4. Seychelles
West Africa	1. Nigeria 2. Ghana 3. Cote d'Ivoire 4. Senegal 5. Mali	1. SierraLeone Nigeria 2. Ghana 3. Cote d'Ivoire 4. Guinea	1. Sierra Leone 2. Niger 3. Nigeria 4. Cote d'Ivoire 5. Ghana	1. Nigeria 2. Mali 3. Ghana 4. Senegal 5. Cote d'Ivoire	1. Ghana 2. Cape Verde 3. Mali	1. Guinea 2. Ghana 3. Cote d'Ivoire 4. Senegal Burkina Faso
Central Africa	1. DR Congo 2. Cameroon 3. Gabon	1. Gabon 2. DR Congo	1. DR Congo 2. Cameroon	1. DR Congo 2. Chad	1. Gabon 2. Cameroon 3. DR Congo	1. Central African Republic 2. DR Congo 3. Cameroon
Southern Africa	1. South Africa 2. Angola 3. Zambia 4. Zimbabwe	1. Mozambique 2. South Africa 3. Zambia 4. Namibia	1. Madagascar 2. South Africa 3. Mozambique 4. Malawi	1. South Africa 2. Zambia 3. Madagascar 4. Namibia	1. Mauritius 2. Botswana 3. South Africa 4. Zambia	1. Zimbabwe 2. Madagascar 3. Malawi 4. Sao Tome and Principe

NDCs, and other strategies, plans, and programs on the focus countries. More so, the approach for estimating CSA investment potential was with a consultation with a wide range of stakeholders to assess the status of climate finance sources and programs.

Experts were drawn from relevant government Ministries, stakeholders in the CSA sector, CSA technology providers, think tanks, civil society organizations (CSOs), Climate Bonds Initiative, and organizations like Africa Renewable Energy Initiative and other policy research institutions to take part in semi-structured face to face interviews. Further, national consultative and validation workshops were conducted with private and public sector players such as the UNFCCC, WB, AfDB, IFC, and other donors. To ensure a bottom-up approach, market players – early-stage, mid-stage, and matured companies – were consulted in developing priority actions for the implementation of CSA technologies. In every stakeholder consultation, a checklist was developed against which information was collected. The information collected was both qualitative and quantitative. To triangulate and complement the empirical findings, the results were further discussed with limited expert stakeholders to assess their opinions and thoughts.

CSA Financial Potential and Barriers Associated with CSA Financing

Financing Potential for Climate-Smart Agriculture in Sub-Saharan Africa

To assess climate smart potential, many adoption studies generally rely on the agricultural practices and opportunities that can be utilized by the private sector (Tesfaye et al. 2017; Pauw and Pegels 2013; Zougmoré et al. 2018; Atteridge and Dzebo 2015; Intellecap 2010; Pauw 2014; Pauw et al. 2015). The evidence generated is qualitative that is rather weak failing to constitute a strong business case to complement the adaptation finance gap in the context of developing countries. It is important to note that the failure to have an argument based on quantitative potential in pursuing private sector engagement in climate-smart agriculture contributes to the adaptation paradox. It is true there are insightful estimates on the business potential of climate-smart agriculture but that is mostly within high level and political contexts. However, agricultural vulnerability is essentially within the local contexts. To engage the local private sector and institutions in implementing the adaptation needs, there is an absolute urgency to create awareness on the underlying climate-smart agriculture potential, investment trends, and agricultural practices with the maximum returns.

Table 3 presents quantified climate investment potential among 14 countries in SSA. It can be observed that Ethiopia has the greatest climate-smart agriculture investment potential of USD 26.4 million. It is followed by Nigeria, Kenya, Madagascar, and Ghana at 17.0, 8.9, 8.4, and 5.5 million USD, respectively. The rest of the countries have a CSA investment potential between the range of 0.3 and 2 million

Table 3 Climate-smart agriculture investment potential

S. No.	Country	Climate-smart agriculture finance potential (USD millions)
1	Ethiopia	26, 400
2	Nigeria	17, 028
3	Kenya	8,910
4	Madagascar	8,360
5	Ghana	5,510
6	Rwanda	2,200
7	Senegal	2,092
8	Cameroon	1,800
9	Mozambique	1,760
10	Ivory Coast	1,656
11	Tanzania	1,500
12	Congo	1,563
13	Zambia	392
14	South Africa	NA
Total		**79,171**

up to the year 2030. In the views of the stakeholders, the cumulative investment potential of CSA is about 40%. Specifically, the private sector can tap into addressing climate change vulnerabilities that are tied to agriculture and water creating a win-win scenario for both the investors and farmers. The figures in Table 3 support the hypothesis that the CSA investment potential is quantifiable and varies across various countries in Africa. According to Quantum that ranks African countries based on Investment Index based on: economic growth, risk factor, business environment, demographic, social capital, and liquidity factor, South Africa, Kenya, and Ethiopia emerged among the top ten (Quantum Global 2018). Essentially, the private sector can single out these three countries on climate-smart sectors.

Climate financing comes from various sources, multilateral and bilateral, public and private, and possibly alternative sources such as remittances (Bendandi and Pauw 2016). According to the OECD (2018)), public climate finance from developed to developing countries in 2017 was at USD 56.7 billion that was a 17% increase from the previous year. Public finance can be used to unlock additional climate funding especially from the private sources that would generally increase the domestic revenue base through proportionate increase of agricultural finance. Despite the continued flow of public financing into developing countries to promote climate adaptation, its documentation as climate-smart agricultural potential for private sector exploration remains unclear.

Table 4 shows that the highest CSA investment is evenly spread among the top five highest receivers. Zambia has the highest investment at 29.4 followed by Madagascar with 28.1, Tanzania 23.4, Nigeria 19.8, and Rwanda 17.31. However, it should be noted that the inflow of multilateral funds in form of adaptation fund supply is 75% grant, 20% concessional loan, and 5% equity. Consequently,

Table 4 Investment trends in climate-smart agriculture in the shortlisted countries

S. No.	Country	Contributions to climate-smart agriculture per year in USD millions
1.	Zambia	29.40
2.	Madagascar	28.06
3.	Tanzania	23.42
4.	Nigeria	19.78
5.	Rwanda	17.31
6.	Ethiopia	17.26
7.	Mozambique	15.77
8.	Senegal	14.96
9.	Cote D'Ivore	13.32
10.	Cameroon	9.67
11.	South Africa	7.99
12	Ghana	7.06
13	Kenya	5.57
14	Congo	4.26
Total		**213.83**

classifying countries according to those with the highest grant inflows that are leaned to agriculture and water sectors, we have Tanzania, Zambia, Ethiopia, Madagascar, and Mozambique.

On average, there is a 98% funding gap in climate-smart agriculture. Out of which, 15–20% can be met by multilateral funds and investments from local governments, financial institutions, and private investors. Arguably, Sub-Saharan Africa will be the global destination for climate financing.

Table 5 identifies some key potential opportunities that could be explored by the private sector across Sub-Saharan Africa. In general, there are investment opportunities in integrated pest and disease control, soil fertility and water management, adoption of new agriculture technologies and practices in addition to farm diversification.

Barriers Hindering the Private Sector from Investing in CSA in Africa

According to the World Bank (2016) observations, investment alone will not be effective in promoting sustainable agriculture. The investments will be rendered ineffective by other existing barriers that are important to untangle. Neglecting uncertainties and failing to factor in nonfinancial and financial-related barriers while making financial decisions may result in wrong investment models that have detrimental effects on both the investor and farmers. Among the many difficulties experienced in trying to close in the financing gap, there is need to reflect on three of the most common barriers: financial, policy and regulatory, government and institutional barriers. Collectively, these barriers directly or indirectly result to income and liquidity variability mostly among the majority of the agriculture sector players – smallholder farmers.

Table 5 CSA implementation measures with the maximum potential for investment by the private sector

S. No.	Country	Climate-smart agriculture interventions with the maximum potential to attract private investment
1	Ethiopia	(a) Periodical application of biofertilizers (b) The precise application of fertilizer
2	Nigeria	(a) Agriculture-based research and development
3	Kenya	(a) Legume-based feeds for dairy cows (b) Organic manure composting and distribution (c) Crop rotation techniques
4	Madagascar	(a) Application of multi-hazard early warning systems and pest control (b) Integrated water resources management specifically in arid areas (c) Large-scale adoption of resilient agriculture
5	Ghana	(a) Agronomic-based support in soil and water conservation techniques (b) Enhancing agricultural productivity (c) Widespread better use of quality fertilizer among the smallholder farmers (d) Promoting agricultural diversification to boost income generation (e) Foster adoption of agriculture-based technologies in water management and small-scale irrigation
6	Rwanda	(a) Soil management in wetlands (b) Multiple mechanisms in pest and disease control (c) Adoption of green manure including crop biomass
7	Senegal	(a) Adoption of sustainable land management technologies (b) Enhancing the adoption of agriculture insurance policies (c) The wide reach of climate-based information (d) Trigger a market base for crop and forest products
8	Cameroon	N/A
9	Mozambique	(a) Adoption of drought-resistant crop varieties (b) Use of integrated pest control and organic manure (c) Management of crop-based biomass (d) Sustainable water management (e) Diversification of sources of livelihoods
10	Cote D'Ivore	(a) Improvement in agricultural production technologies (b) Introduction of agricultural produce storage facilities (c) Popularizing climate-resilient crop varieties
11	Tanzania	(a) Management of soil fertility and extension of agriculture services (b) Introduction of in situ water harvesting techniques and agriculture-based insurance policies (c) Adoption of high yields drought-resistant seed varieties and crop diversification (d) Use of inter alia CSA and widespread knowledge
12	Congo	N/A
13	Zambia	(a) Widespread adoption of drought-resistant crops and agroforestry. (b) Increased biomass capacity. (c) Encouraging fire management and adoption of integrated pest and disease control.
14	South Africa	(a) Adoption of conservation agriculture and better cropping practices. (b) Diversification of farm activities. (c) Livestock and pasture management .

Evidence by Africa Climate Week shows that over 65% of African countries have started their implementation of the Nationally Determined Contributions (Africa Climate Week 2019). Moreover, 80% of the surveyed firms have attained substantial mileage in the adoption and implementation of climate change adaptation measures. However, the survey also established that more than half of African countries face problems in mobilizing both national and international funds. Further on, the Africa Climate Week study found out that over 75% of the surveyed countries did not have an efficient financing strategy with an additional 67% lacking agriculture-based financial instruments. Conclusively, despite the investment potential associated with CSA in Sub-Saharan Africa, it is worthwhile to note that access to climate finance at scale presents a major setback. Divulging more information on financial factors that have large uncertainty along with other intertwining obstacles will be of high value for enhancing investment decisions and policy advocacy. Some of the challenges facing the continent when it comes to CSA financing includes:

  (i)   Overlapping policies on climate change that are weakly enforced.
 (ii)   Inadequate and unstructured provision for climate funding in the respective country's national budgets.
(iii)   Insufficiency in terms of government capacity to satisfy the required standards and procedures needed in developing viable projects and bureaucratic funding processes.
 (iv)   Inadequate knowledge and awareness on the sources of climate finances among the stakeholders in addition to constrained private sector engagement.
  (v)   Lack of appreciation that climate change is both a developmental and environmental concern leading to a silo approach that impairs financing and problem-solving.

Looking more closely at the CSA barriers, most of the stakeholders agreed that financial-related barriers were the greatest hindrance towards unlocking the potential in climate-smart agriculture by the private sector. Secondly, about 80% of the stakeholders interviewed expressed that policy and regulatory setbacks were holding back the private sector from being involved in undertaking climate change interventions in Africa. Additionally, stakeholders highlighted that economic-, infrastructure-, and institutional-related constraints prevented the private sector interest in CSA. It was important to disaggregate the various financial, regulatory, and governance barriers.

Financing products and instruments for climate-smart projects financing from the local commercial banks do not only have high-interest rates and collateral pledges but also short lending tenures. Lending interest rates for the local banks across Africa is between 18% and 20% and in most cases require 100% collateral for one to acquire an agricultural loan. The local commercial banks view climate-smart projects as highly risky due to their limited experience in the newly emerged sector. Arguably, the inadequate data on risk-profile data for climate-smart projects contribute to the financier's views. Also, the target short-term payback period of between 1 and 3 years set for climate-smart projects does not fit with the loan tenors.

Unlike countries such as Bangladesh that experienced a large uptake of climate-smart approaches due to a huge number of microfinance institutions, in SSA, the microlending infrastructure is poorly developed. Microfinancing in the majority of the African countries is still at its infancy, and product sales targeting climate-smart agriculture is virtually nonexistent.

There are conspicuously limited national funds that have been mobilized and distributed at low costs to the private sector. National funds aimed at financing climate-smart agriculture could be of great help in invoking private sector participation with the high costs attracted by commercial bank alternations. However, in the majority of the African countries, national funds for smart agriculture are inactive or are lacking.

Under the current situation, CSA financing in Sub-Saharan African is based on donor funding which is mostly prohibitive due to high mainstreaming and upscaling costs. The import and export of capital in SSA countries are both bureaucratic-lengthy and costly. Climate-smart entrepreneurs and start-ups that are the engine behind innovation and efficient use of resources lack capital. Much of the funding into early-stage business models and start-ups is through private equity, venture capital, or angel investor community that lack the pool of resources necessary to satisfy all sectors. Innovators end up competing against each other dwindling each other's chances of survival. Measures such as promoting blended finance, introducing guarantee funds, encouraging fiscal incentives, establishing accessible local climate funds, and developing friendly investment policies will be key in overcoming the financial barriers and stimulating investor interest.

Key players in smart agriculture need policies that recognize and support the implementation of CSA practices. Investor-based services in smart agriculture like risk insurance and safety nets need considerable policy support. However, in the majority of the SSA countries, several CSA policy loopholes impair the actualization of the action plans. There is a problem with CSA coordination and mainstreaming into the general public. Therefore, the expenditure and planning systems are blurred both at the local, national, and regional levels. It will be critical to strengthening existing synergies to enhance food security programs. The absence of effective policies and regulations discourages lending and creates obstacles to the flow of cash to agriculture. For instance, lack of appreciation by the government of the agriculture economic and market potential lead to ignored subsidies that discourage the development of private sector-based solutions into enhancing climate adaptation.

Financial institutions have limited knowledge on climate-smart projects thereby hindering their investment interest. Both banks and microfinance lack the understanding of the operations of agriculture smart and continuously demonstrate an experience deficit. In such instances, the financial institutions need training on agriculture smart technologies to tailor-make suitable business models and financing options. Circulation of information will be necessary to get rid of the high-risk perception by the lending institutions. Key stakeholders, among them government ministries, nongovernmental organizations, and farmers cooperatives have a knowledge gap on CSA limiting its uptake. It is important for the government to be aware of the agriculture potential in order to promote the development of other sectors that are not necessarily related to agriculture but are indirectly vital for its development, such as the infrastructure and communication networks.

Conclusion

Modern day agriculture has to meet increased food demand due to burgeoning population and evolving diets amidst dwindling crop yield, diminishing natural resources, and constrained biodiversity. Worse is that the continuously warming climate is greatly undermining agricultural productivity with disastrous effects on land, crops, and farmers. Fortunately, adoption of climate-smart agriculture can be a significant part of solving the environmental crisis of climate change. Through sustainable agriculture, there is the capacity to increase agricultural productivity hence increasing incomes through built and adaptive farming resilience. However, this is not possible without substantial increase in the amount of climate-smart investment that will increase the access to finance. There exist huge financial gaps in CSA investment due to the perceptions of high risks and low profitability. Robust financial investment from the private sector can greatly accelerate the adoption of climate-smart agriculture leading to societal gains of poverty reduction by supporting the global food system. Through quantifiable evidence, this chapter has practically provided a translation of the investment potential into investor "language" with the sole objective of invoking private sector interest. The work carried out as the basis of this chapter has gone a step further to identify the maximum potential areas within the smart agriculture in respective countries with a conclusion of disaggregating the sector barriers.

Sub-Saharan Africa has a substantial climate finance investment potential for the private sector. Countries need to promote low carbon development, resource use efficiency, and resilience building in their development strategies and policies. International collaborations should seek to promote regional capacity building in accessing climate finance to promote sustainable development. Multilateral and national climate financing mechanisms should be based on a country's commitment to climate change adaptation. There is a need to strengthen the regulatory environment by creating effective policies and subsidizing the private sector investment to spur adaptation action. SSA needs to foster regional and cross-border collaborations to enhance an integrated approach towards climate change-related issues. National governments need to harness the innovative capacity by raising capital for the private sector that is driving climate investments. Channels to facilitate climate finance into cities and urban areas where there is access to a greater number of people is necessary to greatly reduce the poverty levels.

The contribution of this work is on the conceptual clarifications of the CSA investment potential for the private sector. However, it fails to distinguish the opportunities as either for the domestic or/and international private sector. Again, the chapter outlines climate-smart agriculture potential at a national level. To greatly elicit the private sector, there is need to further breakdown the potential into local contexts. More so, future research can explore the financial potential of each of the identified agricultural practices either regionally or nationally. Further, to objectively attract private investment into climate financing in developing countries, there is need to clearly define the short- and long-term investment activities that can be key for the financial institutions decision-making. The identified limitations create inadequacies in precisely defining the role of the private sector in their increasing engagement on climate finance.

National governments in Sub-Saharan Africa are supporting low carbon and climate-resilient development through local budget allocations, commitments to international programs and strategies that alone cannot achieve the sustainable development goals. They can further enhance their adaptation action through partnering with the private sector to alleviate the funding gap as a result of broadening public debt crises and increasing climate finance needs. This is by creating an enabling environment both in terms of policy, regulations, and infrastructure.

References

Africa Climate Week (2019) Africa climate week ends with calls for investments to tackle climate change – United Nations Sustainable Development. Retrieved from https://www.un.org/sustaina bledevelopment/blog/2019/04/africa-climate-week-ends-with-calls-for-investments-to-tackle-climate-change/

Atteridge A, Dzebo A (2015) When does private finance count as climate finance? Accounting for private contributions towards international pledges. SEI discussion brief. Stockholm

Barkhordarian A, Von Storch H, Bhend J (2012) The expectation of future precipitation change over the Mediterranean region is different from what we observe. Clim Dyn 40(1–2):225–244. https://doi.org/10.1007/s00382-012-1497-7

Barnard J, Manyire H, Tambi E, Bangali S (2015) Barriers to scaling up/out climate smart agriculture and strategies to enhance adoption in Africa. Retrieved from Forum for Agricultural Research in Africa, Accra, Ghana website https://www.nepad.org/publication/barriers-scaling-upout-climate-smart-agriculture-and-strategies-enhance-adoption

Bendandi B, Pauw P (2016) Remittances for adaptation: an 'alternative source' of international climate finance? In: Migration, risk management and climate change: evidence and policy responses, pp 195–211. https://doi.org/10.1007/978-3-319-42922-9_10

Branca G, Tennigkeit T, Mann W, Lipper L (2012) Identifying opportunities for climate-smart agriculture investments in Africa. FAO

Buah SS, Ibrahim H, Derigubah M, Kuzie M, Segtaa JV, Bayala J et al (2017) Tillage and fertilizer effect on maize and soybean yields in the Guinea savanna zone of Ghana. Agric Food Secur 6 (1). https://doi.org/10.1186/s40066-017-0094-8

Christiansen L, Ray AD, Smith JB, Haites E (2012) Accessing international funding for climate change adaptation. Climate and sustainable development. UNEP Risø Centre on Energy, Roskilde

ECOWAS is in High Level Forum of Climate Smart Agriculture Stakeholders Intervention Framework for the Development of Climate Smart Agriculture under the West Agricultural Policy (ECOWAP/CAADP) implementation With technical facilitation by: In partnership with: Accelerating the implementation of ECOWAP/CAADP High Level Forum of Climate Smart Agriculture Stakeholders in West Africa Bamako (Mali), June 15–18, 2015 Intervention Framework for the Development of Climate Smart Agriculture under the West Africa Regional Agricultural Policy (ECOWAP/CAADP) implementation Process

Elrafy M (2009) Impact of climate change: vulnerability and adaptation of coastal areas. Report of the Arab Forum for Environment and Development

FAO (2012) Identifying opportunities for climate-smart agriculture investments in Africa. Retrieved from http://www.fao.org/docrep/015/an112e/an112e00.pdf

FAO (2013) Climate smart agriculture source book. Retrieved from Food and Agriculture Organization of the United Nations website http://www.fao.org/3/i3325e/i3325e.pdf

FAO, IFAD and WFP (2014) The State of Food Insecurity in the World 2014. Strengthening the enabling environment for food security and nutrition. Rome, FAO.

Giller KE, Witter E, Corbeels M, Tittonell P (2009) Conservation agriculture and smallholder farming in Africa: the heretics' view. Field Crop Res 114(1):23–34. https://doi.org/10.1016/j.fcr.2009.06.017

Hoerling M, Hurrell J, Eischeid J, Phillips A (2006) Detection and attribution of twentieth-century northern and southern African rainfall change. J Clim 19(16):3989–4008. https://doi.org/10.1175/jcli3842.1

Intellecap (2010) Opportunities for private sector engagement in urban climate change resilience building. Author, Hyderabad

IPCC (2014) Climate change 2014: impacts, adaptation and vulnerability. IPCCWGIIAR5 technical summary. Retrieved from http://ipccwg2.gov/AR5/images/uploads/WGIIAR5-TS_FGDall.pdf

Jalloh A, Roy-Macauley H, Sereme P (2012) Major agro-ecosystems of West and Central Africa: brief description, species richness, management, environmental limitations and concerns. Agric Ecosyst Environ 157:5–16. https://doi.org/10.1016/j.agee.2011.11.019

Lobell DB, Bänziger M, Magorokosho C, Vivek B (2011) Nonlinear heat effects on African maize as evidenced by historical yield trials. Nat Clim Chang 1(1):42–45. https://doi.org/10.1038/nclimate1043

Maguza-Tembo F, Edriss A, Mangisoni J (2017) Determinants of climate smart agriculture technology adoption in the drought prone districts of Malawi using a multivariate probit analysis. Asian J Agric Ext Econ Soc 16(3):1–12. https://doi.org/10.9734/ajaees/2017/32489

Mapfumo P, Onyango M, Honkponou SK, El Mzouri EH, Githeko A, Rabeharisoa L et al (2015) Pathways to transformational change in the face of climate impacts: an analytical framework. Clim Dev 9(5):439–451. https://doi.org/10.1080/17565529.2015.1040365

Mbow C, Smith P, Skole D, Duguma L, Bustamante M (2014) Achieving mitigation and adaptation to climate change through sustainable agroforestry practices in Africa. Curr Opin Environ Sustain 6:8–14. https://doi.org/10.1016/j.cosust.2013.09.002

Mhlanga N, Blalock G, Christy R (2010) Understanding foreign direct investment in the southern African development community: an analysis based on project-level data. Agric Econ 41(3–4):337–347

Milder J, Majanen T, Scherr S (2011) Performance and potential of conservation agriculture for climate change adaptation and mitigation in Sub-Saharan Africa: an assessment of WWF and CARE projects in support of the WWF-CARE alliance's rural futures initiative. Retrieved from https//:Barriers-to-scaling-up-out-CSA-in-Africa.pdf

Nciizah AD, Wakindiki II (2015) Climate smart agriculture: achievements and prospects in Africa. J Geosci Environ Protection 03(06):99–105. https://doi.org/10.4236/gep.2015.36016

OECD (2018) Public climate finance to developing countries is rising – OECD. Retrieved from https://www.oecd.org/environment/public-climate-finance-to-developing-countries-is-rising.htm

Partey ST, Thevathasan NV, Zougmoré RB, Preziosi RF (2016) Improving maize production through nitrogen supply from ten rarely-used organic resources in Ghana. Agrofor Syst. https://doi.org/10.1007/s10457-016-0035-8

Pauw WP (2014) Not a panacea: private-sector engagement in adaptation and adaptation finance in developing countries. Clim Pol 15(5):583–603. https://doi.org/10.1080/14693062.2014.953906

Pauw P, Pegels A (2013) Private sector engagement in climate change adaptation in least developed countries: an exploration. Clim Dev 5(4):257–267. https://doi.org/10.1080/17565529.2013.826130

Pauw WP, Klein RJ, Vellinga P, Biermann F (2015) Private finance for adaptation: do private realities meet public ambitions? Clim Chang 134(4):489–503. https://doi.org/10.1007/s10584-015-1539-3

Quantum Global (2018) Africa investment index 2018. Retrieved from Renew website https://quantumglobalgroup.com/wp-content/uploads/2017/04/Africa_Investment_Index_April_2017_18.04.2017Final_Curves.pdf

Radhouane L (2013) Climate change impacts on North African countries and on some Tunisian economic sectors. J Agric Environ Int Dev 107:101–113

Sanou J, Bationo BA, Barry S, Nabie LD, Bayala J, Zougmore R (2016) Combining soil fertilization, cropping systems and improved varieties to minimize climate risks on farming productivity in northern region of Burkina Faso. Agric Food Secur 5(1). https://doi.org/10.1186/s40066-016-0067-3

SER (2011) Advies Ontwikkeling door duurzaam ondernemen. Sociaal Economische Raad, Den Haag

Surminski S (2013) Private-sector adaptation to climate risk. Nat Clim Chang 3(11):943–945. https://doi.org/10.1038/nclimate2040

Sylla MB, Gaye AT, Jenkins GS (2012) On the fine-scale topography regulating changes in atmospheric hydrological cycle and extreme rainfall over West Africa in a regional climate model projections. Int J Geophys 2012:1–15. https://doi.org/10.1155/2012/981649

Teklewold H, Mekonnen A, Kohlin G, Di Falco S (2017) Does adoption of multiple climate-smart practices improve farmers' climate resilience? Empirical evidence from the Nile basin of Ethiopia. Clim Change Econ 08(01):1750001. https://doi.org/10.1142/s2010007817500014

Tesfaye K, Kassie M, Cairns JE, Michael M, Stirling C, Abate T et al (2017) Potential for scaling up climate smart agricultural practices: examples from sub-Saharan Africa. Clim Change Manage:185–203. https://doi.org/10.1007/978-3-319-49520-0_12

Thornton PK, Jones PG, Owiyo T, Kruska RL, Herrero M, Orindi V, Bhadwal S, Kristjanson P, Notenbaert A, Bekele N, Omolo A (2008) Climate change and poverty in Africa: Mapping hotspots of vulnerability. Afr J Agr Res Econ 311-2016–5524, 21. https://doi.org/10.22004/ag.econ.56966

Torquebiau E (2015) Whither landscapes? Compiling requirements of the landscape approach. In: Minang P et al (eds) Climate-smart landscapes. ICRAF, Nairobi

Tran NL, Rañola RF, Ole Sander B, Reiner W, Nguyen DT, Nong NK (2019) Determinants of adoption of climate-smart agriculture technologies in rice production in Vietnam. Int J Clim Change Strategies Manag 12(2):238–256. https://doi.org/10.1108/ijccsm-01-2019-0003

UFCC (2007) Investment and financial flows UNFCCC to address climate change: an update Bonn. Technical Paper FCCC/ TP/2008/7

UFCC (2011) Report of the conference of the parties on its sixteenth session, held in Cancun from 29 November to 10 December 2010

UNDESA (2014) Urbanization prospects. The 2014 revision. Highlights. United Nations, New York

UNEP FI (2009) The materiality of climate change. How finance copes with the ticking clock. United Nations Environment Programme Finance Initiative, Geneva

United Nations, Department of Economic and Social Affairs, Population Division (2019) World urbanization prospects: The 2018 revision (ST/ESA/SER.A/420). United Nations, New York

Wambugu C, Place F, Franzel S (2011) Research, development and scaling-up the adoption of fodder shrub innovations in East Africa. Int J Agr Sustain 9(1):100–109. https://doi.org/10.3763/ijas.2010.0562

Williams T, Mul M, Cofie O, Kinyangi J, Zougmoré R, Wamukoya G (2015) Climate smart agriculture in the African context background paper feeding Africa conference, 21–23 Oct 2015

World Bank (2015) World development indicators 2015. Author, Washington, DC

World Bank (2016) Making climate finance work in agriculture. Retrieved from http://documents.worldbank.org/curated/en/986961467721999165/pdf/ACS19080-REVISED-OUO-9-Making-Climate-Finance-Work-in-Agriculture-Final-Version.pdf

Zougmoré R, Traoré AS, Mbodj Y (eds.) (2015) Overview of the scientific, political and financial landscape of climate-smart agriculture in West Africa. Working Paper No. 118. CGIAR research program on climate change, agriculture and food security. Retrieved from http://www.ccafs.cgiar.org

Zougmoré RB, Partey ST, Ouédraogo M, Torquebiau E, Campbell BM (2018) Facing climate variability in sub-Saharan Africa: analysis of climate-smart agriculture opportunities to manage climate-related risks. Cahiers Agric 27(3):34001. https://doi.org/10.1051/cagri/2018019

10

Access to Water Resources and Household Vulnerability to Malaria in the Okavango Delta, Botswana

M. R. Motsholapheko and B. N. Ngwenya

Contents

Abstract

Malaria is a persistent health risk for most rural communities in tropical wetlands of developing countries, particularly in the advent of climate change. This chapter assesses household access to water resources, livelihood assets, and vulnerability to malaria in the Okavango Delta of north-western Botswana. Data were obtained from a cross-sectional survey of 355 households, key informant interviews, PRA-

M. R. Motsholapheko (✉)
Water Resources Management Program, Okavango Research Institute, University of Botswana, Maun, Botswana
e-mail: rmoseki@ub.ac.bw

B. N. Ngwenya
Ecosystems Services Program, Okavango Research Institute, University of Botswana, Maun, Botswana
e-mail: bntombi@ub.ac.bw

based focus group discussions (FDGs), interviews with experts in various related fields, PRA workshop participant interviews, and literature review. There was high access to natural capital, and most households engaged in nature-based livelihood activities. Access to resources determined type of livelihood activities that households engaged in. However, there was no association between household exposure and/or susceptibility, and type of livelihood activities pursued by households. Household vulnerability to malaria was higher in remote and rural locations than in urban neighborhoods. Malaria prevention and vulnerability aversion programs need to be coupled with improvements in housing and well-being in the Okavango Delta and similar wetlands.

Keywords

Livelihoods · Malaria · Vulnerability · Wetlands

Introduction

In developing countries, access to natural resources largely determines human livelihoods and general well-being of households in many rural communities (Rebelo et al. 2010). Due to lack of access to various forms of capital including financial, physical, human, and social capital, and because of general poverty the dependence on natural resources has been increasing in wetland communities around the world. This is because wetlands act as sources of various forms of ecosystem services that support livelihoods. At the same time, wetlands provide suitable habitats for disease-carrying organisms and therefore remain vector- and water-borne diseases endemic areas. Some vector-borne diseases such as malaria account for millions of human morbidity and deaths annually. For instance, the World Health Organization (WHO) reported that in 2018 malaria accounted for 228 million morbidity cases and 405,000 deaths globally, 94% of which were in Africa (WHO 2019). Due to climate variability and change, the prevalence and virulence of vector-borne diseases (VBDs), including malaria, may increase (IPCC 2007). The challenges associated with malaria outbreaks include deaths of breadwinners in mostly poor households, which largely depend on the labor of their able-bodied members. Given that such mortalities often result in increased suffering of mostly women, children (WHO 2019), and other vulnerable groups, the fight against malaria requires concerted efforts on the part of governments, NGOs, communities, and other stakeholders to enhance access to prevention and treatment. Prevalence of malaria in tropical and subtropical regions including Africa, continues to overwhelm existing measures for prevention and treatment. This leads scientists and policymakers to assess existing forms of prevention and treatment with a view to developing effective means for combating malaria.

Various approaches have been adopted particularly in the prevention and treatment of malaria including (a) distribution of free insecticide-treated mosquito nets (ITNs) in rural communities, (b) use of preventive antimalarial medicines,

(c) interior and exterior spraying, (d) awareness creation, and (e) educational campaigns. Most of these efforts have been supported by governments, NGOs, and global organizations including the World Health Organization. As a result, there is a general decline in the prevalence of malaria in developing countries. In some countries, where malaria prevalence had been recorded and often led to deaths, malaria has been successfully eradicated.

Botswana is one of the 21 countries which were identified by the WHO for malaria elimination by 2020 (WHO 2019). Botswana is also among eight southern African countries which formed the Elimination-8 partnership aimed at eradicating cross-border malaria transmission by 2030 through collaborations and synchronization of policies and initiatives. At the national level, a rapid notification and response strategy was developed to eliminate malaria in hotspot areas, including the Ngamiland District in the Okavango Delta in Botswana (Chihanga et al. 2016).

Despite all these efforts, malaria remains pervasive in wetlands of developing countries, including Botswana, particularly among poor households. This calls for additional efforts to determine the general vulnerability of communities to malaria to improve general knowledge on factors that contribute to its persistence. This chapter is aimed at improving knowledge on household vulnerability to malaria in the Okavango Delta and similar areas around the world. Specifically, the chapter assesses (a) household access to water resources and (b) household vulnerability to malaria in Shakawe and Ngarange villages and the surrounding areas. The chapter is arranged as follows: the next section discusses the conceptual framework used, followed by environmental and socioeconomic context, research approach and data collection, the results on access to capital and household vulnerability, and finally, the section on discussion, conclusion, and implications for malaria policies in the Okavango Delta and other wetlands in similar socioeconomic and environmental conditions.

Conceptual Framework

This research was informed by the sustainable livelihood framework supplemented by the socio-ecological framework. These frameworks were selected for their wide use in microeconomic analyses on access to resources, rural livelihoods, and household vulnerability to shocks in developing countries. The frameworks differ in their approach to the household vulnerability context; the former views livelihood shocks as exogenous factors that are beyond household control, whereas the latter posits that shocks and household livelihoods are interlinked within the broad human–environment interactions (Berkes and Folke 1998; Ellis 2000). Both frameworks were influenced by the changing view on vulnerability as not just the biophysical nature of the shock but the state of the affected social unit as well (Wisner et al. 2004).

The sustainable livelihood framework provides an understanding of the broad concept of a livelihood, particularly the vulnerability context within a livelihood is constructed. In the sustainable livelihood framework, a livelihood can be defined as

"...comprising of assets (natural, physical, human, financial, and social capital), the activities, and the access to these (mediated by institutions and social relations) that together determine the living gained by the individual or household" Ellis (2000, p. 30). The sustainable livelihood framework has some limitations because it does not cater for changes over time (Ellis 2000). It is nonhistoric because it takes current household access to resources as given, without looking at the origins and possible causes of current access conditions (Small 2007). Furthermore, it neither provides for tracing historical factors leading to current institutional structure, nor for understanding the nature of interactions between actors brought together by various interests and complying to set rules, norms, and practices (O'Laughlin 2002). For instance, it does not clearly show the role of the private sector as part of the organizations that mediate access to forms of capital. Furthermore, it does not provide adequate guidance on the analysis of institutions and the behavior of actors involved.

Another observation is that the vulnerability context in the sustainable livelihood framework does not consider the full context of shocks, trends, and seasonality, partly because it does not clearly show the position of the household itself. Although it considers both the negative and positive aspects of these processes, it does not clearly show how negative livelihood outcomes can exacerbate the impacts of shocks in the vulnerability context. This may be because it construes shocks as exogenous factors and therefore beyond household control (Ellis 2000). The sustainable livelihood framework is an integrating analytical tool which was never intended to be used in isolation (Farrington et al. 1999). The sustainable livelihood framework was supplemented with the socio-ecological framework to overcome some of its limitations. The socio-ecological framework was useful in locating the household construction of a livelihood within the broad human–environment interactions in which it occurs. In other words, it helped clarify the broad livelihood vulnerability context in the Okavango Delta. The socio-ecological complex or framework emanated from classical human ecology in the 1930s (Berkes and Folke 1998). It was influenced by the concept of resilience which has now become important for understanding complex relationships between human and other ecological systems (Folke 2001). The socio-ecological framework has been modified and used for analyzing human-environment interactions in developing countries and providing the basis for some contemporary vulnerability frameworks (Arntzen 1989; Adger 2006). This framework analyses rural development in terms of clusters of elements and interactions among clusters, which finally lead to a sustainable society or community (Arntzen 1989; Berkes and Folke 1998). It distinguishes four elements that characterize the socio-ecological linkages and interactions. These are population, organization, the environment, and technology at the micro-scale, with influencing factors at the macro-scales (regional, national, and global) (Arntzen 1989; Berkes and Folke 1998).

Understood in the context of this framework, the household as a human construct (illustrated in Fig. 1 below), and the livelihood as a subset of the household system, fall within the broad human–environment system.

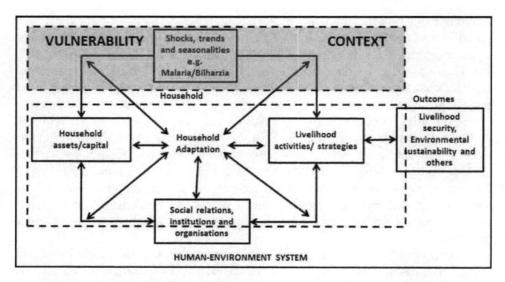

Fig. 1 Sustainable livelihood framework. (Source: Adapted from Ellis (2000, p. 30) with insights from Berkes and Folke (1998, p. 15))

Therefore, the impacts of a shock such as malaria, on a household livelihood and the household responses are part of the interface between the human and environment subsystems. In other words, it is acknowledged that livelihood shocks such as malaria are exogenous to a household because they fall beyond household control as posited by Ellis (2000). However, the effect of their impacts falls within household control depending on household exposure, susceptibility, and capacity to adapt. Therefore, such effects are part of the household vulnerability complex made up of the social (human) and biophysical (or ecological) aspects (Marshall et al. 2009). In the case of malaria endemicity in the Okavango Delta, household adaptation which is the inverse of vulnerability can therefore be viewed as the process and actions within a livelihood system occurring or being undertaken to positively influence or be influenced by livelihood security and environmental sustainability. This position is informed by the view that a livelihood is sustainable if it can adapt, be resilient, and therefore less vulnerable to shocks while not degrading the natural resource base (Scoones 1998).

Household adaptation to a livelihood shock such as malaria may occur at various dynamic levels influencing or being influenced by livelihood security and or environmental sustainability. It may also reduce the impacts of a shock on both livelihood assets and household livelihood activities, as well as the impacts of the institutions, social relations, and organizations, which modify or mediate access to household assets. Household actions to reduce vulnerability to malaria may also affect the setting up of livelihood activities with influence on the type of activity and location within which they may be undertaken. Of course, in case of maladaptation which has a positive effect in increasing vulnerability, the influence will be negative, leading to livelihood insecurity and environmental degradation. Of interest in this chapter is the household access to assets in the process of constructing livelihood activities and how such access exposes and makes households susceptible to malaria as a health risk. The chapter focuses on the vulnerability context as illustrated in Fig. 1 above.

Okavango Delta: Environmental and Socioeconomic Context

The Okavango Delta is an alluvial fan and a vast wetland well-renowned for its biodiversity, species richness, and prominence as a Ramsar Site, and 1000th UNESCO World Heritage Site in north-western Botswana. Due to its richness in flora and fauna, the Okavango Delta enhances dynamic livelihood systems within a semiarid environment of the southern African Kalahari plains. The Okavango Delta and related floodplains, at an altitude of 900 m above sea-level, within a low-physiographic zone given the generally flat terrain, and daily temperatures ranging from 12 °C to 40 °C provide a suitable habitat for the anopheles mosquito (*Anopheles arabiensis*). Rural communities found in many settlements around the Delta fringes as they depend on the natural resources in the form of flora such as papyrus, riverine reeds, grasses, and others, as well as faunal species including fish, birds, and small game.

The specific research sites were the villages of Shakawe and Ngarange in the Okavango sub-District of Ngamiland District. Ngamiland West, a Ministry of Health designated area which largely covers the Okavango sub-District, is among the poorest regions in the country with 47.3% of people below the poverty datum line compared to the national figure of 20.7% (Statistics Botswana 2011). Ngamiland West is a malaria-endemic area due to the high persistence of this disease in many of its villages. HIV/AIDS prevalence in the Ngamiland District is also high (18%), compared to the national average of 17.6% (Central Statistics Office 2009) (Fig. 2).

Shakawe and Ngarange villages are in an area commonly referred to as the panhandle, which is made up of the Okavango River as it enters Botswana from the Kavango Region of Namibia. Shakawe village has a population of 6693 people and Ngarange has 998 people. The population structures of these two villages (Fig. 3) depict a wide base in the young age groups compared to advanced age groups; the patterns that are typical of population structures found in developing countries.

However, there appears to be an anomaly in that the population pyramids in these villages have a narrowed base particularly in the infant cohorts and under-five age group, which is indicative of low fertility and possibly high mortality in these age groups. This is a logical observation in that due to high prevalence of HIV/AIDS, women may tend to postpone their engagement in sexual activity, or totally abstain from sex to avoid repeat infections and/or opt for effective birth prevention methods (Lewis et al. 2004). Due to illness and the impacts of opportunistic diseases, HIV/

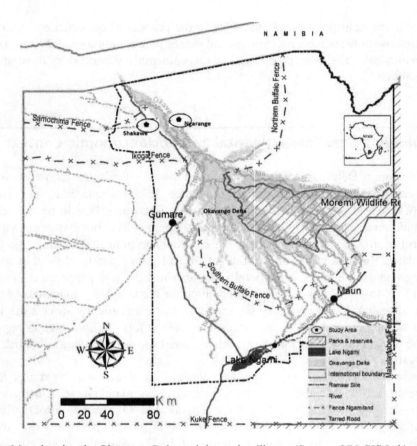

Fig. 2 Map showing the Okavango Delta and the study villages. (Source: ORI GIS Lab)

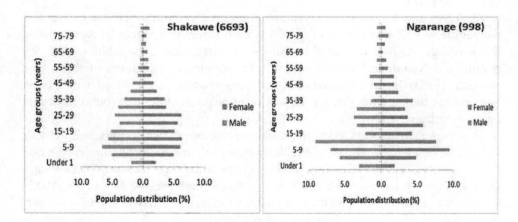

Fig. 3 Population structure in Shakawe and Ngarange villages

AIDS also has adverse effects on fertility (Ross et al. 1999). Infant and under-five mortality is generally higher in communities with high HIV/AIDS prevalence than in those with low prevalence and this may have been contributed to the narrow pyramid base in these two villages. From the foregoing, households in Shakawe and surrounding villages including Ngarange may be vulnerable to other diseases including malaria.

Shakawe has been identified as a malaria hotspot in Botswana mainly due to the high number of confirmed cases in this village. Ngarange has been observed to have low confirmed cases compared to other settlements in the same area. These two villages were selected to understand this apparent contrast and general dynamics surrounding household vulnerability to malaria in the Okavango Panhandle and the Okavango Delta in general.

Research Approach and Data Collection

A convergent parallel mixed-method approach was adopted; it involved the concurrent use of qualitative and quantitative methods of data collection and analysis. This approach made it possible to exhaustively obtain information from the various methods used and substantively confirm outcomes through triangulation. It was also noninterventional given that no further action was taken in treating those infected with malaria; affected individuals received treatment from local health clinics in the respective villages.

Secondary data were obtained through desktop searches in the World Wide Web and from the review of published and unpublished sources including journal articles, global and regional health (and other) organizations' reports, and the national health policy and strategy documents. Primary data were obtained, (a) through a participatory rural appraisal (PRA) methodology which included key informant interviews and PRA-based focus group discussions, (b) through a household survey and informal discussions with community members, and (c) from interviews with various stakeholders including officers in local health clinics, the district office in Maun, the sub-District main office in Gumare, and from the Ministry of Health head office in Gaborone.

Participatory Rural Appraisal (PRA) Workshop

A PRA workshop was held on June 25–27, 2014 in Shakawe with the aim to create project awareness, stakeholders' involvement, and participation in the project, and to collect baseline data on key areas that required further scientific enquiry. In the latter aim, the researchers were able to document community knowledge on livelihoods, the social, environmental, and climate change factors that influence household vulnerability to vector-borne diseases including malaria. Stakeholders including traditional leaders, traditional health practitioners, representatives of various village committees, modern medical practitioners, government health officers, and nongovernmental organizations' (NGOs) representatives were invited to the workshop.

The criteria for selecting other members of the community included (a) being adult women and men above the age of 18 years, (b) having lived in the locality for 10 years or more, and (c) being knowledgeable on vector-borne diseases. A total of 40 people, 25 (63%) from Shakawe, and 15 (38%) from Ngarange were invited; the total consisted of 13 (32%) female and 27 (68%) male. Given the diversity of stakeholders and level of representation, the PRA workshop provided opportunities for interviews with the attendants individually, as key informants, discussants in focus group discussions, and as participants in a plenary.

Household Survey

The household survey was undertaken following a stratified sampling of households. Information on malaria occurrence was first obtained from the local clinics in the villages of Shakawe and Ngarange. Using such information, village wards were identified and differentiated according to high, medium, and low number of malaria cases. Village wards were then randomly selected from the broad categories of high, medium, and low malaria cases. In Shakawe, a total of seven out of fourteen wards were selected. Although the same stratification criteria of high, medium, and low malaria cases were used in Ngarange, given the small size of the village, all wards were selected. After the selection of wards in both villages, all households in the selected wards were listed. From a list of households in all the selected wards, households were selected proportionate to size of ward. In Shakawe, a total of 877 households were listed in the seven wards, whereas in Ngarange 300 households were listed in the four wards, therefore some overall total of 1177 households were listed in these two villages. The sample size in the two villages was calculated at 95% confidence interval with margin of error of plus/minus five (+/−5). An overall sample of 355 households (266 in Shakawe and 89 in Ngarange) was selected in the two villages (Table 1).

Table 1 Number of households listed and sampled in various wards of Shakawe and Ngarange villages

Village	Ward	Listed households ($N = 1177$)	Sampled households ($n = 355$)
Shakawe ($n = 266$)	Mabudutsa	29	9
	Matomo	55	17
	Ukusi	83	25
	Ndumbakatadi	163	49
	Saoshoko	178	54
	Diseta	140	42
	Gauxa	229	70
Ngarange ($n = 89$)	Modubana	103	30
	Newtown	48	14
	Mukumbe	107	33
	Sekandeko	42	12

A questionnaire comprising structured and semistructured questions were developed. It had sections on (a) household demographic and socioeconomic characteristics, (b) types of livelihood activities, (c) access to and ownership of assets or capital, (d) malaria prevention and coping strategies and other sociocultural variables. Data on household characteristics, livelihood types, and malaria prevention and coping strategies were captured through open- and close-ended questions. Data on access to capital were obtained through several proxy indicators. For example, access to financial capital was measured using questions on whether (or not) a household had access to cash, credit, and insurance, and access to physical capital was measured through questions on ownership of assets such the plowing equipment, vehicles, telecommunication equipment, and so on.

Various forms of capital were measured using different units; therefore, composite unitless indices similar to the UNDP's human development index were developed to consolidate the various measurement units. Data on household adaptive capacity, susceptibility, and exposure were also captured and transformed into unitless indices. Vulnerability was then calculated using a composite index, being a function of exposure less adaptive capacity, multiplied by susceptibility. Details of the techniques used are described in detail in Hahn et al. (2009), Motsholapheko et al. (2011), and UNDP (1990), so they will not be repeated in this chapter. The questionnaire was administered through face-to-face interviews with household heads, their spouses, and adult representatives. The survey data collection was undertaken in the period October 27 to November 6, 2015, in the two villages of Shakawe and Ngarange.

Access to Water Resources in Shakawe and Ngarange

Results from the survey (Fig. 4) reflected generally similar patterns of access to capital in the two villages. Access to natural capital was higher than that of other capital types in both Ngarange and Shakawe. However, specific differences can be identified; households in Ngarange had slightly higher access to natural capital than those in Shakawe; the index values were 0.67 and 0.61, respectively. Worth noting is that access index values are all above 0.5 and edge toward the value of one which is the maximum. This result is not surprising given that Ngarange has a small population and the village is remotely located on the eastern side of the river which has vast floodplains than in the western escarpment where Shakawe is located. The proportions of households who stated that they had unimpeded access to land, water, forest, and aquatic resources were higher in Ngarange, ranging from 63% to 75%, compared to Shakawe where these proportions ranged from 56% to 72%. In wetlands, such as the Okavango Delta, rural communities depend on natural resources to make a living.

Ngarange households had slightly higher access to financial capital than their counterparts in Shakawe with index values of 0.5 and 0.36, respectively. This was a counter-intuitive result given that Shakawe, as the main commercial center in the area, is more urbanized than Ngarange. A further assessment of household characteristics and livelihood activities revealed that most households in Shakawe were

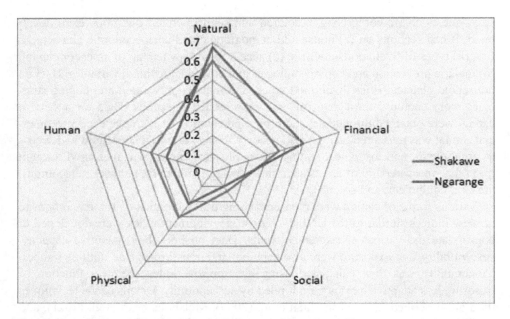

Fig. 4 Household access to resources in Shakawe and Ngarange villages

de jure female-headed (57%) whereas male-headed households accounted for 33% and the rest were de facto female-headed at 10% (De jure female-headed households are those that are directly headed by females, whereas de facto female are those headed by females in the absence of the substantive male head (Motsholapheko et al. 2011).). Conversely, most households in Ngarange were male-headed (44%), the remaining being *de jure* (25%) and de facto (31%) female-headed. Interviews with experts on livelihoods and social development in the area revealed that although Shakawe is more urbanized than most settlements around it, the village could have experienced retarded development related to ownership of land. Large portions of undeveloped land in the village belong to a few individuals whose capacity to develop may either be limited or depends on self-interest and monopolistic tendencies. A further scrutiny of the above indices for Ngarange (0.5) and Shakawe (0.36) indicates a generally low access to financial capital in both villages, given that households have access values of up to half or less than the maximum value of one.

Access to physical, human, and social capital was higher in Shakawe than in Ngarange. Index values were 0.31 and 0.21 for physical capital, 0.34 and 0.25 for human capital, and 0.13 against 0.1 for social capital in Shakawe and Ngarange, respectively. A key observation here is that although Shakawe village households have a slight edge over those in Ngarange, all the above values are very low ranging from 0.26 to 0.36 which reveals very low access compared to a maximum value of one. In terms of ranking, access to natural capital would be the highest followed by financial, human, physical, and social capital in that order. Social capital was least accessed by households in both villages. Key informant interviews have revealed that access to social capital in these villages was adversely affected by loss of culture, breakdown in family ties, individualism, and general lack of cooperation within the village community.

Household Vulnerability to Malaria in Shakawe and Ngarange

Results from PRA-based key informant interviews (Table 2) revealed a general awareness among interviewees that malaria exposure was dominant among some social groups than others within their communities.

Most key informant interviewees (78%) stated that under-five children were the most exposed to malaria whereas safari operators were the least exposed, being stated by 35% of the interviewees. According to key informants, the top five most affected social groups were under-five children, primary school, commercial fishers, pregnant women, hook and line fishers, and river reed/grass harvesters. Commercial fishers and pregnant women were stated by an equal proportion of key informants at 52%. The various social groups stated above are linked to the types of livelihood activities they engage in, as well as their levels of possible susceptibility to malaria given their physical condition. For example, under-five children and pregnant women are known to be more susceptible to most diseases due to their physical condition (Bates et al. 2004).

Key informant interviewees also perceived that some livelihood activities such as commercial fishing and river grass harvesting were high-risk activities for malaria exposure; they were stated by 62.5% and 50% of the interviewees, respectively (Table 3). The top five livelihood activities stated by interviewees as likely to expose households and community members to malaria were commercial fishing, grass harvesting, reed harvesting, livestock herding, hook and line fishing, and arable farming.

Table 2 Key informants' perceptions on social groups and vulnerability to malaria

Social group	Proportion of key informants % ($n = 40$)	Top five social groups
Children <5 years	78	1. Children <5
Primary school children	58	2. Primary school children
Pregnant women	52	3. Commercial fishers and pregnant women
Commercial fishers	52	
Hook and line fishers	50	4. Hook and line fishers
Reed/grass harvesters	48	5. Reed/grass harvesters
Polers	40	
Cattle herders	38	
Safari operators	35	

Table 3 Proportions of key informants and their perceptions on level of risk of malaria infection and related livelihood activities

Livelihood activities	Proportion of key informants (%) by perception of level of risk ($n = 40$)			
	High	Medium	Low	Don't know
Commercial fishing	62.5	22.5	10	5
Grass harvesting	50	20	15	15
Reed harvesting	47.5	17.5	25	10
Livestock herding	45	7.5	32.5	15
Hook and line	42.5	17.5	32.5	7.5
Arable farming	42.5	22.5	25	10
Tourist poling	40	15	22.5	22.5
Basket Fishing	37.5	22.5	27.5	12.5
Tswii harvesting	37.5	12.5	32.5	17.5
Tour guiding	35	15	15	35
Molapo farming	30	12.5	10	47.5
Basket weaving	20	12.5	50	17.5

Basket weaving was perceived by 50% of interviewees as less likely to expose households and community members to malaria. Furthermore, *molapo* farming was one of the activities viewed as less likely to expose households and community members to malaria being stated by 30% interviewees. However, most interviewees (47.5) indicated that they did not know the likely level of exposure to malaria for *molapo* farming households and community members. This could be due to insufficient knowledge on *molapo* farming among Shakawe and Ngarange communities. Studies have indicated that *molapo* farming is not widely practiced in these communities compared to their counterparts in the mid- and lower parts of the Okavango Delta.

Results from the socioeconomic survey (Fig. 5 below) indicated that the top five livelihood activities undertaken by households in Shakawe were river grass harvesting in which 60% of households were involved, reed harvesting (58%), dryland/rain-fed arable farming (49%), livestock farming (35%), and *tswii* (waterlily or Nymphaea tuber) harvesting (30%).

These results corroborated key informant interview results on four of the five livelihood activities. Although hook-and-line fishing and commercial fishing were indicated as possible high-risk activities, the socioeconomic survey indicated that these livelihood activities were undertaken by a low proportion of households at 15% and 4.8%, respectively.

Household Exposure, Susceptibility, and Adaptive Capacity

Survey results (Fig. 6) revealed that households in Ngarange village had higher exposure to malaria (0.5) than their counterparts in Shakawe (0.34). These results are quite realistic in that Ngarange village is located near a vast floodplain where water flow is limited compared to Shakawe, which is located along an escarpment.

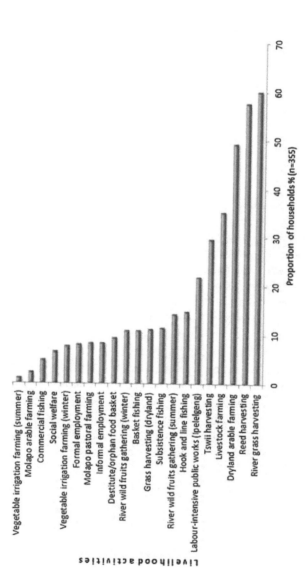

Fig. 5 Proportion of households and livelihood activities in Shakawe and Ngarange. Note: Other livelihood activities accounted for 0.3–0.6% of households

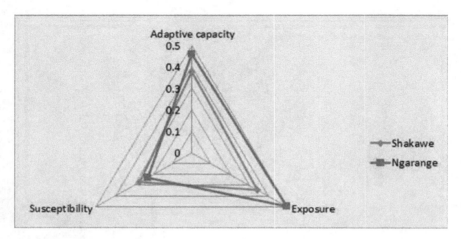

Fig. 6 Village-level susceptibility, adaptive capacity and exposure to malaria

Ngarange households also had higher adaptive capacity (0.46) than those in Shakawe (0.38). Worth noting is that household exposure (0.5) to malaria in Ngarange was higher than adaptive capacity (0.46) meaning that within the village, households could not adequately prevent malaria transmission. In Shakawe, house-holds had higher adaptive capacity (0.38) than exposure (0.34) meaning that they could prevent malaria transmission.

In both villages, the overall household susceptibility to malaria was lower than both adaptive capacity and exposure. Ngarange had a slightly lower susceptibility value (0.23) than Shakawe (0.27), which indicates that households in both villages had almost the same level of susceptibility to malaria. A further assessment of household exposure, adaptive capacity, and susceptibility within these two villages indicates that households in five out of seven wards in Shakawe had higher adaptive capacity than exposure to malaria. The five wards of Diseta, Matomo, Ndumbakatadi, Mabudutsa II, and Saoshoko had adaptive capacity that exceeded exposure values, whereas in the two wards of Gauxa and Ukusi, the exposure values of 0.39 and 0.44 were higher than the respective adaptive capacity values of 0.35 and 0.28 (Fig. 7).

This indicates that households in Gauxa and Ukusi wards of Shakawe were not adequately adaptable to malaria than their counterparts in the other five wards. Household susceptibility in to malaria in four of the seven wards of Shakawe was higher than both exposure and adaptive capacity. The susceptibility values in these wards were 0.65 or higher compared to both exposure and adaptive capacity values that ranged from 0.27 to 0.44.

In Ngarange, household exposure was higher than both adaptive capacity and susceptibility to malaria in all wards except in Mukumbe Ward, which had a susceptibility value of 0.63 that was higher than both adaptive capacity and expo-sure. These results indicate that households in all the wards in Ngarange are not adaptable and as such vulnerable to malaria, particularly Mukumbe Ward which had higher susceptibility than the other wards.

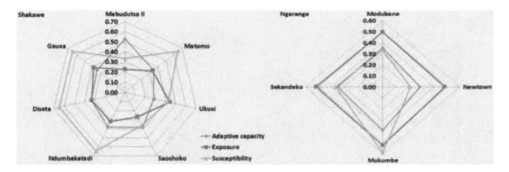

Fig. 7 Household adaptive capacity, susceptibility and exposure to malaria in Shakawe and Ngarange

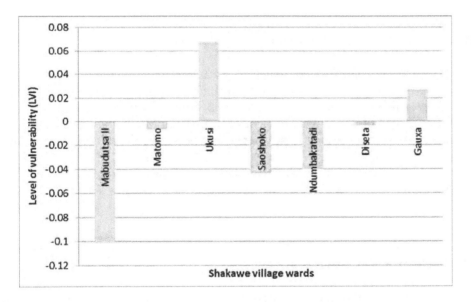

Fig. 8 Level of household vulnerability (LVI) by ward in Shakawe

The overall livelihood vulnerability indices for these villages were -0.0092 for Shakawe and 0.0088 for Ngarange, which confirms the above suggestion that households in Ngarange were not adequately adaptable and therefore vulnerable to malaria whereas their counterparts in Shakawe were not vulnerable. A further analysis indicates that households in Mabudutsa II ward of Shakawe had the lowest vulnerability score of -0.101 (Fig. 8), which indicates that they could withstand the impacts of malaria much better than their counterparts in other wards.

The other four wards of Saoshoko (-0.043), Ndumbakatadi (-0.04), Matomo (-0.007), and Diseta (-0.003) also had low vulnerability scores. The two wards of Ukusi (0.068) and Gauxa (0.027) were vulnerable to malaria given that they had positive values. The vulnerability scores for the two wards of Matomo and Diseta

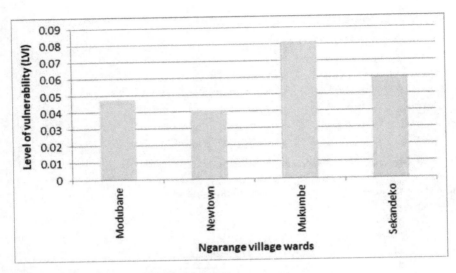

Fig. 9 Level of household vulnerability (LVI) by ward in Shakawe

were close to the value of zero, which indicates that even though these two wards were not vulnerable, any slight decline in adaptive capacity or increase in exposure to malaria may render households in these two wards vulnerable to malaria.

In Ngarange, all wards were vulnerable to malaria, Mukumbe Ward having the highest vulnerability value of 0.08 whereas Newtown Ward had the lowest vulnerability at 0.04 (Fig. 9).

A further analysis indicated that households in the two most vulnerable wards of Mukumbe and Sekandeko in Ngarange were the most susceptible and least adaptable to malaria.

Discussion and Conclusion

This chapter assessed household access to water resources as well as vulnerability to malaria in the Okavango Delta. The results indicate that household access to natural capital was much higher than that of other types of capital. Households mostly engaged in nature-based livelihood activities such as rain-fed arable agriculture, livestock, and the harvesting of aquatic reed, river grass, and *tswii* (nymphaea). Most Ngarange households engaged in nature-based livelihoods than their counterparts in Shakawe village, which was relatively more urbanized than Ngarange. Social capital was the least accessed of all the five forms of capital. It has also been demonstrated that households in Ngarange were more vulnerable to malaria than their counterparts in Shakawe and that this vulnerability emanated from high-household exposure mainly due to low capacity to adapt to malaria than their counterparts in Shakawe.

Due to high access to natural capital compared to other capital forms, households engaged in nature-based livelihood activities. This is typical of most rural communities in developing countries, particularly those residing in wetlands of sub-Saharan

Africa (Babulo et al. 2008; Rebelo et al. 2010; Angelsen et al. 2014). As earlier indicated in the conceptual framework, households' choice of livelihood activity depends on access to capital such that the most accessed type of capital may determine the main livelihood activity in any livelihood portfolio (Ellis 2000; Berkes and Folke 1998).

Access to social capital was lower than that of other capital forms contrary to expectation that this form of capital is higher in rural communities than in urban settings. Key informants attributed this low access to social capital to loss of culture, breakdown in family ties, a rise in individualism and general lack of cooperation within the village community. Low access to social capital may have adverse impacts on the well-being of households given that cultural diversity, social cohesion, and network are critical components of social sustainability (Munasinghe 2000).

Households in Ngarange were found to be more vulnerable to malaria than their counterparts in Shakawe. This vulnerability could be attributed to high exposure to malaria partly due to type of housing, and household practices relating to malaria prevention. Some studies have posited that malaria transmission is related to type of housing, and with increased urbanization and change of housing structure malaria transmission is likely to decrease in sub-Saharan Africa (Saugeon et al. 2009; Parnell and Walawege 2011). Household livelihood activities (such as agriculture) and practices have also been found to be confounding factors for malaria transmission, in Africa and Southeast Asia even in urban communities (Klinkenberg et al. 2008). Effectively, malaria transmission directly relates to type of housing, preventive measures, and practices relating to mosquito bites avoidance (Tusting et al. 2016).

Vulnerable groups such as pregnant women, under-five children, and the elderly require targeted programs with emphasis on availing the means to prevent malaria transmission. There is a need for long-term, sustained prevention, and control to reduce and/or eliminate malaria in the Okavango Delta and Botswana, in general.

References

Adger WN (2006) Vulnerability. Glob Environ Chang 16:268–281

Angelsen A, Jagger P, Babigumira R, Belcher B, Hogarth NJ, Bauch S, Rner JB, Smith-Hall C, Wunder S (2014) Environmental income and rural livelihoods: a global-comparative analysis. World Dev 64:S12–S28

Arntzen J (1989) Environmental pressure and adaptation in rural Botswana. PhD thesis, Vrije University of Amsterdam, Amsterdam

Babulo B, Muys B, Nega F, Tollens E, Nyssen J, Deckers J, Mathijs E (2008) Household livelihood strategies and forest dependence in the highlands of Tigray, northern Ethiopia. Agric Syst 98:147–155

Bates I, Fenton C, Gruber J, Lalloo D, Lara AM, Squire SB, Theobald S, Thomson R, Tolhurst R (2004) Vulnerability to malaria, tuberculosis, and HIV/AIDS infection and disease. Part 1: determinants operating at individual and household level. Lancet Infect Dis 4:267–277

Berkes F, Folke C (1998) Linking social and ecological systems for resilience and sustainability. In: Berkes F, Folke C, Colding J (eds) Linking social and ecological systems: management practices and social mechanisms for building resilience. Cambridge University Press, Cambridge

Central Statistics Office (2009) 2008 Botswana AIDS impact survey III statistical report. Central Statistics Office, Gaborone

Chihanga S, Haque U, Chanda E, Mosweunyane T, Moakofhi K, Jibril HB, Motlaleng M, Zhang W, Glass GE (2016) Malaria elimination in Botswana, 2012–2014: achievements and challenges. Parasit Vectors 9:99. https://doi.org/10.1186/s13071-016-1382-z

Ellis F (2000) Rural livelihood diversity in developing countries. Oxford University Press, Oxford

Farrington J, Carney D, Ashley C, Turton C (1999) Sustainable livelihoods in practice: early applications of concepts in rural areas. Natural resources perspectives paper number 42. Overseas Development Institute, London

Folke C (2001) Socio-ecological resilience and behavioral responses. http://www.beijer.kva.se/publications/pdf-archive/Disc155.pdf. Accessed 07 June 2018

Hahn MB, Riederer AM, Foster SO (2009) The livelihood vulnerability index: a pragmatic approach to assessing risks from climate change variability and change – a case study in Mozambique. Glob Environ Chang 19:74–88

Intergovernmental Panel on Climate Change [IPCC] (2007) Climate change 2007: impacts, adaptation and vulnerability. Contribution of working group II to the fourth assessment report of the Intergovernmental Panel on Climate Change. Cambridge University Press, Cambridge, UK

Klinkenberg E, McCall PJ, Wilson MD, Amerasinghe FP, Donnelly MJ (2008) Impact of urban agriculture on malaria vectors in Accra, Ghana. Malar J 7:151

Lewis JJC, Ronsmans C, Ezeh A, Gregson S (2004) The population impact of HIV on fertility in sub-Saharan Africa. AIDS 18(2):S35–S43

Marshall NA, Marshall PA, Tamelander J, Obura D, Malleret-King D, Cinner JE (2009) A framework for social adaptation to climate change: sustaining tropical coastal communities and industries. IUCN, Gland

Motsholapheko MR, Kgathi DL, Vanderpost C (2011) Rural livelihoods and household adaptation to extreme flooding in the Okavango Delta, Botswana. Physics and Chemistry of the Earth 36:984–995

Munasinghe M (2000) Development, equity and sustainability (DES) in the context of climate change. Guidance Papers on the Cross Cutting Issues of the Third Assessment Report of the IPCC, 69–90

O'Laughlin B (2002) Proletarianization, agency and changing rural livelihoods: forced labor and resistance in colonial Mozambique. J South Afr Stud 28(2):511–530

Parnell S, Walawege R (2011) Sub-Saharan African urbanization and global environmental change. Glob Environ Chang 21(1):12–20

Rebelo LM, McCartney M, Finlayson C (2010) Wetlands of sub-Saharan Africa: distribution and contribution of agriculture to livelihoods. Wetl Ecol Manag 18(5):557–572

Ross A, Morgan D, Lubega R, Carpenter LM, Mayanja B, Whitworth JA (1999) Reduced fertility associated with HIV: the contribution of pre-existing subfertility. AIDS 13:2133–2141

Saugeon C, Baldet T, Akogbeto M, Henry MC (2009) Will climate and demography have a major impact on malaria in sub-Saharan Africa in the next 20 years? Med Trop 69(2):203–207

Scoones I (1998) Sustainable rural livelihoods: a framework for analysis. IDS working paper 72. Institute of Development Studies, London

Small L (2007) The sustainable rural livelihoods approach: a critical review. Can J Dev Stud 28:27–38

Statistics Botswana (2011) Botswana core welfare indicators (poverty) survey of 2009/10 preliminary results. Statistics Botswana, Gaborone

Tusting LS, Rek J, Arinaitwe E, Staedke SG, Kamya MR, Cano J, Bottomley C, Johnston D, Dorsey G, Lindsay SW, Lines J (2016) Why is malaria associated with poverty? Findings from a cohort study in rural Uganda. Infect Dis Poverty 5:78

United Nations Development Program [UNDP] (1990) Human development report. UNDP, New York

Wisner B, Blaikie P, Cannon T, Davis I (2004) At risk: natural hazards, people's vulnerability and disasters, 2nd edn. Routledge, London

World Health Organization [WHO] (2019) World malaria report 2019. World Health Organization, Geneva

11

ECMWF Subseasonal to Seasonal Precipitation Forecast for use as a Climate Adaptation Tool Over Nigeria

Ugbah Paul Akeh, Steve Woolnough and Olumide A. Olaniyan

Contents

Abstract

Farmers in most parts of Africa and Asia still practice subsistence farming which relies minly on seasonal rainfall for Agricultural production. A timely and accurate prediction of the rainfall onset, cessation, expected rainfall amount, and its intra-seasonal variability is very likely to reduce losses and risk of extreme weather as well as maximize agricultural output to ensure food security.

U. P. Akeh (✉) · O. A. Olaniyan
National Weather Forecasting and Climate Research Centre, Nigerian Meteorological Agency, Abuja, Nigeria
e-mail: paulugbah@gmail.com

S. Woolnough
Department of Meteorology, University of Reading, Reading, UK
e-mail: s.j.woolnough@reading.ac.uk

Based on this, a study was carried out to evaluate the performance of the European Centre for Medium-range Weather Forecast (ECMWF) numerical Weather Prediction Model and its Subseasonal to Seasonal (S2S) precipitation forecast to ascertain its usefulness as a climate change adaptation tool over Nigeria. Observed daily and monthly CHIRPS reanalysis precipitation amount and the ECMWF sub-seasonal weekly precipitation forecast data for the period 1995–2015 was used. The forecast and observed precipitation were analyzed from May to September while El Nino and La Nina years were identified using the Oceanic Nino Index. Skill of the forecast was determined from standard metrics: Bias, Root Mean Square Error (RMSE), and Anomaly Correlation Coefficient (ACC).

The Bias, RMSE, and ACC scores reveal that the ECMWF model is capable of predicting precipitation over Southern Nigeria, with the best skill at one week lead time and poorest skills at lead time of 4 weeks. Results also show that the model is more reliable during El Nino years than La-Nina. However, some improvement in the model by ECMWF can give better results and make this tool a more dependable tool for disaster risk preparedness, reduction and prevention of possible damages and losses from extreme rainfall during the wet season, thus enhancing climate change adaptation.

Keywords

Evaluation · Subseasonal to seasonal (S2S) · Forecast · Metrics · Skill · Climate change adaptation · Ocean Nino Index · ElNino · LaNina · Precipitation amount

Introduction

Background

Weather and climate affect man in diverse ways which greatly impact agriculture, food security, housing, transportation, health, engineering structures, water resources, military operations, oil exploration and exploitation, environment, livelihoods, etc. (Ugbah 2016). These impacts are usually associated with extreme weather and seasonal hazards such as tropical cyclones, heavy rainfall, flooding, etc., which occur more frequently (IPCC 2018). For these reasons, scientists have continued to invest great time and resources in trying to understand and predict the processes that lead to changes in the atmosphere associated with these hazards. One of such investment is the international project known as the subseasonal to seasonal (S2S) prediction project set up in 2013 by the World Meteorological Organization (WMO), World Weather Research program (WRWP), and World Research and Climate program (WRCP) to provide a common platform for sharing of data and knowledge between countries, National meteorological Agencies, and researchers which will enhance our understanding of seasonal and subseasonal variability and also help to improve the skill and accuracy of long range forecast (weeks–months) and likely extreme events in different parts of the world (Takaya 2015).

In most parts of Asia and Africa and in developing countries such as Nigeria, farmers still practice subsistence farming which relies mainly on seasonal rainfall for Agricultural production (Siegmund et al. 2015; Hansen 2002). Farming, out of all human activities, is the most weather-dependent (Hanson 2002). A timely and accurate prediction of the onset, cessation, expected rainfall of the monsoon season, and its intra-seasonal variability is very likely to reduce losses and risk of extreme weather (Vitart et al. 2016) as well as maximize agricultural output to ensure food security. This is one of the purposes for which the S2S project was set up.

The time range for subseasonal forecast lies between 2 weeks and 60 days (Vitart et al. 2016) within the rainy season and it is meant to "bridge the gap between the medium range forecast" (at most 2 weeks) and seasonal forecast (3–6 months) [Vitart et al. 2016; Molteni et al. 2016]. The Madden Julian Oscillation (MJO), among other factors such as El-Nino Southern Oscillation (ENSO), stratospheric influence, Ice and snow cover, soil moisture, ocean conditions, and tele-connection between the tropics and extra tropics, is said to be the main source of predictability for sub-seasonal time range (Vitart et al. 2016; Inness and Dorling 2010); this may not be unconnected with the fact that MJO plays a significant role in initiating or terminating equatorially trapped waves (Kelvin waves) and Rossby waves (Maclachlan et al. 2015) which influence ENSO phenomena. The choice of the ECMWF forecast rather than others is based on the high skill displayed by the model in simulating MJO over the tropics and extra tropics, and the continuous improvement in skill of predicting subseasonal variability and events up to lead time of 10 days, since 2002. This is attributed to the improvement in the ECMWF model characteristics (resolution, initial conditions, physics, etc.) made possible because of the advancement in information technology which has led to the production of better computers with higher computational capabilities, capacity, and speeds in recent times. This development gives a ray of hope to those who depend on subseasonal to seasonal rainfall for production, planning, decision making, disaster preparedness, and socioeconomic development, thus the need to embrace the S2S project.

This study was done in 2016 with the aim of assessing the skill of ECMWF subseasonal ensemble forecasting system in simulating subseasonal rainfall variability in terms of pattern and amount of rainfall in Nigeria up to lead time of one month specifically over southern Nigeria where annual rainfall is highest and the risk of flooding from excessive rainfall also high. The area is highly threatened by sea level rise especially during the rainy season. This work also investigates how good the S2S model ensemble is, in predicting the likelihood of extreme rainfall that may cause flooding over the region.

The scope of this work is limited to southern Nigeria only where risk of flooding is highest and does not cover the entire country. Also it does not include evaluation of how the model can simulate or predict rainfall onset, cessation, or length of season but rainfall amount and pattern. It is expected that findings from this research will not only help to improve the quality and accuracy of S2S forecast, but will also help in updating the seasonal forecast of Nigerian Meteorological Agency and provide useful information and early warning system for disaster preparedness and risk reduction against extreme weather and climate events such as flooding.

Review of ECMWF Subseasonal to Seasonal Forecasting

In 2002, Frederic Vitart started the ECMWF medium range (monthly) forecast of precipitation as an experimental phase which became fully operational by 2004. It was integrated with the ensemble prediction system in 2008 following results of improved skill in simulation of weekly surface temperature, precipitation, mean sea level, and MJO observed during the period over the extra-tropics (Vitart 2005, 2014; Vitart et al. 2016). The model is run two times in a week to produce a daily forecast in real time for lead period up to 32 days, with a 1 hourly high resolution retrieval, and a reforecast (hindcast) for more than 12 years (Molteni et al. 2016; Vitart et al. 2016). The forecast consists of 51 ensemble members while the reforecast has 11. According to Vitart et al. (2016), it is structured to allow better coupling of the atmosphere (IFS model) and ocean (NEMO model) compared to models used in NOAA and UKMet office. Representation of model physics and physical processes has also been improved in very recent version. Some of the improvements include the increase in frequency of the S2S forecast from every 14 days in 2002 to twice a week from 2013 to date. The horizontal resolution has also evolved from T159 which had a lead time up to 32 days in 2002, through T255, T319, and presently T639 with lead time of up to 46 days. Vertical resolution has also improved from 40 atmospheric levels with top at 10 hPa in earlier versions to 91 atmospheric levels with top at 1 Pa in 2016. Re-forecast size now has 11 ensemble members as against 5 members in earlier versions (Molteni et al. 2016).

Work by Vitart (2005; Vitart et al. 2016) in trying to evaluate the skill of ECMWF subseasonal forecast showed higher weekly scores for lead times greater than 10 days and better skills when compared to persistent forecast and climatology (Vitart 2005; Vitart et al. 2016; Haiden et al. 2015). In another work, verification of the ECMWF subseasonal forecast model was done using metrics such as anomaly **coefficient of correlation (ACC) and root mean square error RMSE and probabilistic skills** (Vitart 2004**).** This approach is very popular and it is in line with the WMO recommendation on verification of deterministic forecast (Inness and Dorling 2010). **Vitart's result showed decreased correlation in the weekly anomalies with increasing lead time; the highest correlation value was observed in the first week forecast.** The skill was found to be greater than skill of the persistence forecast. For the ensemble probability forecast, the relative operative characteristics (ROC) scores which show the ratio between the Hit rate and false alarm were determined using the contingency table. ROC value of 0.5 means "no skill forecast" while values above 0.5 suggest good skill of forecast. The ROC values obtained from Vitarts work showed that forecast of 12–18 days was better than persistence forecast of 5–11 days. His work in 2014 also confirms this, with highest skill for week 1 (0.8) decreasing to 0.6 during week 3 (Vitart 2014). Brier score of ECMWF subseasonal forecast was 0.04 which suggested that model performed better than climatology (Vitart et al. 2016). The result of evaluation of the ECMWF subseasonal forecast showed better performance than other global centers as seen from the studies of Vitart over the extra tropics and consistent with other centers (Haiden et al. 2014). He pointed out that there could be variability in scores from one region to another and so, it would be wrong to conclude that the African region with a different climate, weather and geomorphology, will produce same skill as the extra-tropics without carrying out studies to compare and make valid conclusions, hence the justification for this study over Nigeria.

Li and Robertson (2015) also carried out studies to evaluate precipitation forecast from global ensemble prediction system at submonthly timescales using hind cast data from three global prediction centers (ECMWF, NCEP, and JMA) during May to September for the period 1992–2008. Statistical metrics used were the Correlation of Anomalies (CORA), and the Mean Square Skill Score (MSSS). The results showed good skills from all the three models during the first week compared to lead times of 2–4 weeks. The skill from ECMWF forecast was however better than the other two. Anomaly correlation skill scores were found to be high at lead times 1–4 over the equatorial Pacific while values of 0.2–0.3 were found to be statistically significant over tropical Atlantic at lead 1 only. The skills over land areas were generally poor even at lead time of one week over Africa. The result also showed very good correlation between ENSO and the submonthly prediction with ACC skills for ENSO years higher than neutral years. Biases in mean weekly precipitation forecast were large particularly in the NCEP model over the Atlantic, West Africa, and the Sahel regions.

Studies by Hamill (2012) on verification of the ECMWF probabilistic precipita-tion forecast, TIGGE National Centers for Environmental Prediction (NCEP) model, Canadian model, and UK Met Office (UKMO) forecast models during June–November 2002 to 2009 over the United states of America showed that ECMWF had the best skills with better scores for higher precipitation. These works confirm the leading standard of the ECMWF model forecast which has been chosen for this study.

Description of the Study Areas and Methodology

Study Area

Nigeria is located towards the coast of West Africa on latitude 4.0°N–14.0°N and longitude 2.45°E–15.5°E (Fig. 1). It occupies a landmass of about 924,000 Km^2 and population of about 133 million as at 2006 which grows at a rate of 2.8% (Ayanlade et al. 2013). The study area is the southern part of the country shown in the red box (Fig. 1) defined by latitude 4.0°N–8.0°N and longitude 2.0°E–11.0°E estimated from the figure which is sourced from Eludoyin and Adelekan (2013). It covers an area less than 50% of the country and is bounded in the South by the Atlantic Ocean (Gulf of Guinea). The area is bounded in the East and West by two countries: Cameroun and Benin Republic, respectively.

Climatology and Sources of Subseasonal Predictability in the Region

The study area lies within the mangrove swamp, tropical rainforest, and Guinea Savanna vegetative zones which experience two major seasons: Wet (late February

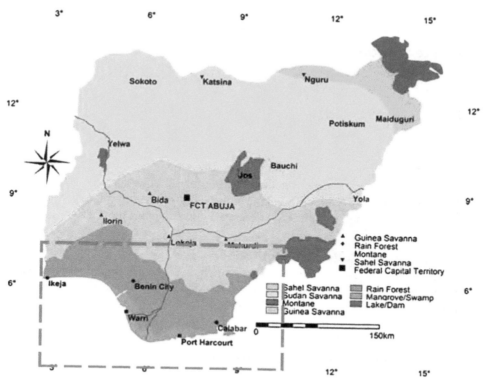

Fig. 1 Map of Nigeria showing study area in rectangular box (Source: Eludoyin and Adelekan 2013, springer publishers)

to November) and the dry (December to mid-February) as reported by Eludoyin and Adelekan (2013). The unequal distribution of rainfall from month to month within these two major seasons further distinguishes the season into long rainy season (March–July), short dry season or August break (Late July–early August), short rainy season (early September–mid-October), and the long dry season (late October–early March) (Chidiezie and Shrikant 2010). The wet season is said to have a bimodal pattern of rainfall and a longer period which lasts about 200–300 days (NiMet 2013, 2014). It is characterized by moist southwesterly winds also known as the tropical maritime air mass (mT) which blow landwards from the Atlantic Ocean and is predominant during the rainy season. The dominant winds during the short period of dry season (Harmattan season) are the dry dusty northeasterly trade winds from the desert also called the tropical continental air mass (cT) which find their way to the South from the North (Eludoyin & Adelekan 2013). Annual rainfall total ranges from 1500 to 3000 mm over most of the South but can reach about 4000 mm over the southeastern part where rainfall is highest (Ayanlade et al. 2013, Chidiezie and Shrikant 2010). Mean air temperatures are in the range of 25.0°–30.0 °C during summer and 20.0–30.0 °C during the harmattan season (Ayanlade et al. 2013)

Research Methods

Study Site Selection and Sampling Methods
The study area covers all locations in southern Nigeria which lie between latitude 4.0 and 8.0 N and longitude 2.0 and 11.0 E which are Bayelsa, Rivers, Akwa-Ibom, Cross-river, Ebonyi, Enugu, Abia, Anambra, Imo, Delta, and Edo states. Others are Ondo, Lagos, Ogun and southern part of Osun, Oyo, Benue, and Kogi states. The selection of the area is intentional for only areas which experience high annual rainfall of 1500 mm and above and prone to high risk of flooding due to the heavy rainfall and proximity to the Atlantic Ocean. The area is also close to rivers Niger and Benue and selected based on three agro-climatic zones: mangrove swamps, tropical rainforest, and Guinea Savanna only. This means that a combination of purposeful and stratified sampling was employed in selecting the study area.

Data Sources and Collection Methods
Secondary data was used for this study and includes forecast and observed precipitation data for 21 years from 1995 to 2015 over Nigeria. This is the period for which reforecast data was available. The forecast data is the ECMWF ensemble reforecast precipitation data in mm/day downloaded from the S2S data website hosted by ECMWF and it is available up to lead time of 40 days from 11 ensemble members. The mean ensemble precipitation forecast data provided at a resolution of $1.5° \times 1.5°$ grids (IRI 2016) was used for this study. The observed rainfall data is the CHIRPS reanalysis of daily precipitation provided at a resolution of 0.05×0.05 grid which is higher than the resolution of the forecast data. It was downloaded from the IRI website earlier discussed and credited to Funk et al. (2014).

No missing data was observed in the data sets as it is likely that any existing missing data may have been filled during reanalysis/homogenization before making the data available for users on the websites.

Weekly precipitation totals and climatological averages of observation and forecast from May 14 to September 24 (20 weeks) for the 21 year period (1995–2015) was extracted and used to study the intra-seasonal changes in precipitation. The choice of the seasonal period (May–September) is based on the enormous rainfall usually experienced during this period. ENSO years within the period were selected based on the Oceanic Nino index (ONI) provided by NOAA.

Data Analysis
Codes were written first, to read all the data files which were in NetCDF format, and then to convert the time variable which was in Julian days in CHIRPS data to the normal calendar dates to enable easy extraction of weekly totals of precipitation from May 14 to September 24. This was then followed by extraction of weekly means and climatological averages of precipitation over southern Nigeria from which temporal plots were made. Time series were plotted as average weekly while the climatological average was determined from the weekly totals averaged over 21 years. Bias and RMSE between forecast and observed precipitation was calculated accordingly.

Anomalies were determined by subtracting the climatological mean from the weekly mean precipitation and the anomalies correlated to determine the ACC skill of the forecast at different lead times. All plots as well as calculations were made using Python.

The ECMWF forecast data, unlike CHIRPS, had a land-sea masked file for measurements over land and ocean. This masked file was used to mask out the ocean precipitation values in order to extract only values over land. The step is taken to isolate the land grid boxes when calculating precipitation averages and totals to enable a fair comparison between the data sets.

The observed precipitation for the previous week was persisted for the following week to produce the one week lead time persistent forecast for the 20 weeks period (May 14–September 24). This was then compared with the observed climatology and forecasts at weekly lead times of one, two, three, and four weeks referred to as lead 1, 2, 3, and 4. The El Nino and La Nina events were identified by averaging the five consecutive Oceanic Nino Index (ONI) values of months AMJ, MJJ, JJA, JAS, ASO, and the average was compared with the threshold that defined an ENSO event, that is, ONI values greater or equal to +0.5 °C define an El Nino while values less or equal to −0.5 is La Nina (NOAA 2016). These months were chosen to be consistent with the season (May–September) for which forecast is made. The flow chart (Fig. 2) summarizes the procedure involved.

Fig. 2 Flow chart of methodology. (Source: Ugbah 2016)

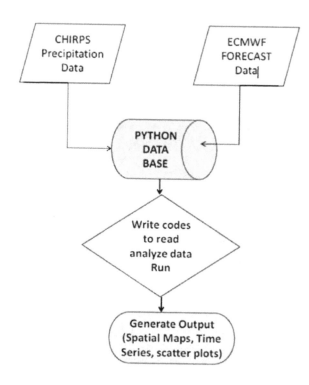

Results and Discussion

Comparison Between Observation and Forecast at Different Lead Times

Weekly totals of the forecast and observed precipitation covering the twenty weeks period (May 14 to September 24) from 1995 to 2015 were plotted to compare the extent to which the observation agrees with or deviates from the forecast at different lead times. The result is shown in Fig. 3.

It can be seen from Fig. 3 that both the forecast and observation follow a similar pattern during the season and depict the two peaks (bimodal) of rainfall in early July and in September. The short period of the little dry season known as "August break" is also shown by the precipitation minimum around late July/August. At initial start dates up to June 11 and towards the end of the season in September, the precipitation values all seem to be in close agreement, having only small differences, but beyond the week starting June 11, the forecasts start to deviate significantly from the observed with the largest differences around the period of the little dry season after which the lines begin to converge until the end of the season (Fig. 3). It can also be seen that the differences in precipitation amount increase with increasing lead

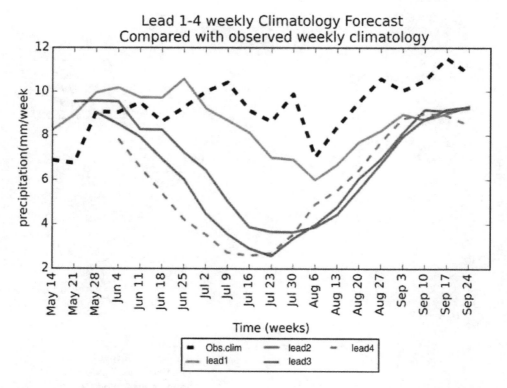

Fig. 3 Forecast at different weekly lead times (wk1-wk4) indicated with colored lines and Observed weekly climatology (black dotted line) compared (Source: Ugbah 2016)

time as depicted by lead 1 forecast (red line) which lies closest to the observation (dotted black line) throughout the season. Forecast of lead 2, 3, and 4 show larger departures from the observed climatology (Fig. 3) with the largest departure seen in the lead 4 forecast (purple dotted line). These large departures possibly arise from the inability of the model to properly simulate the little dry season phenomenon and related subseasonal processes. Correction of these differences is likely to improve the performance of the model. The plot also shows that the predicted precipitation at all the different lead times is lower than the observed implying an underestimation by the model. This can also be corrected to improve the model.

Evaluation of the Forecast Skill: Metrics

It is important to explain some of the different scores that are used for evaluating this forecast as there are different types of metrics used for verification of forecast and the choice of a particular type depends on the objective of the verification. Some of the popular metrics used by scientists and recommended by WMO for use in verification of precipitation forecasts include bias, root mean square error (RMSE), and anomaly correlation coefficient (ACC). These metrics have been selected for this study to provide a common standard for comparing the results with those of other scientist and meteorologist on related works.

Bias

The bias refers to the differences between the forecast and the observation mean. It is a measure of the extent to which the ECMWF model under-estimates or over-estimates precipitation. Bias information is useful in correcting errors in models which are likely to improve its performance (Inness and Dorling 2010). Mathematically it is defined as:

$$\text{Bias} = \sum_{i=1}^{n} \frac{(f - O)}{N}$$

where f = forecast, O = observation, and N = total number of events.

A larger bias shows large differences or disagreement between forecast and observation while smaller values suggest closeness or similarities. Such closeness may not necessarily mean that the forecast is perfect because observations could be wrong and therefore give misleading results. Bias can be positive when forecast is over-estimated or negative when under-estimated.

The biases were computed at the four different lead times and then averaged over the twenty weeks period from May to September (Table 1). Negative biases can be seen for leads of 1, 2, 3, and 4 weeks. The biases increase with lead time from week one (-0.69) to week four (-3.10). This is consistent with findings from Vitart 2004, 2014, and Li and Robertson (2015). At lead times of 2 to 4 weeks, the dry bias is particularly strong from late June through July. This suggests difficulty by the model

Table 1 1995–2015 ECMWF forecast vs observed precipitation scores at weekly lead times

Scores of the ECMWF subseasonal weekly forecast over Southern Nigeria				
Lead time scores	Week 1	Week 2	Week 3	Week 4
BIAS	−0.69	−2.30	−2.7	−3.10
RMSE	1.64	3.33	3.78	3.98

to forecast at a period when reduced rainfall is experienced in the West African Monsoon (WAM) and when the Inter-Tropical Convergence Zone (ITCZ) also called ITD is at its most northward position from the area of study. Such biases may be necessary to correct only if other metrics also provide evidence in support of poor model performance. This is because bias alone is inadequate in evaluating the performance of a model as it only tells only the differences and does not give weight to errors like the RMSE. That is why the use of only one metric is not recommended but a combination of two or more.

Root Mean Square Error (RMSE)

RMSE can be defined as the square root of the sum of the squares of all errors (biases) between a given set of data as shown below (Chai and Draxler 2014).

This metric is good for use in evaluating the performance of a model because it represents normally distributed errors better than uniformly distributed errors common in models (Chai and Draxler 2014). It gives more weight to large errors than smaller errors and has a tendency of showing some bias towards datasets without large errors, but it still remains one of the oldest and most popularly used metric among scientists in model evaluation. Large RMSE values imply large error distribution while smaller values suggest smaller errors. RMSE of zero means that the forecast is without errors or bias.

$$RMSE = \sum_{i=1}^{N} \frac{(f - O)2}{N}$$

where f = forecast, O = observation, N = total number of events.

The results of RMSE also show that errors increased with lead time, and were lowest (1.64) at lead time of one week and highest (3.98) at lead time of four weeks (Table 1).

Anomaly Correlation Coefficient (ACC)

ACC was used to evaluate the potential skill of the forecast. It is considered a potential skill because it is only one metric out of the several others that need to be considered to arrive at the final or perfect skill of a forecast. It may not be exhaustive, but it provides very useful information for assessing model performance and a pointer to the likely overall skill. Correlation between forecast and CHIRPS observed daily precipitation data was done using python program scripts written to incorporate the mathematical formula for ACC defined by Maue and Langland (2014) as:

$$ACC = \frac{\sum_{m=1}^{m} f' * O'}{\sqrt{\left(\sum_{m=1}^{m} (f')2 * \sum_{m=1}^{m} (O')2\right)}}$$

where $f' = fm\text{-}Cm$, is the forecast mean (fm) – climatological mean of forecast (Cm)
$O' = Om\text{-}Cm$, is the mean of observed (Om) – Observed mean Climatology (Cm).

The climatological average values of forecast and observation for the 21 year
period was first calculated and then subtracted from each weekly forecast and observed
precipitation to get the desired result. Correlation values generated from linear regression analysis of these anomalies also produced similar results when compared. This
served as a check for possible computational errors from code written to calculate this
metric. One advantage with this metric is the consideration of errors averaged over a
longer period than for short period averages which are usually highly variable. This
advantage, on the other hand, also constrains the metric in its ability to significantly
reflect changes in the mean values over shorter time ranges.

The results of the correlation for weekly start dates during the season and at lead
times of the 1, 2, 3, and 4 weeks indicated are further discussed in section "Performance of the S2S Model Precipitation Forecast During Rainfall Peaks and Period of
Little Dry Season". There are 20 weekly start dates in the season starting with the
first on May 14, then May 21, May 28, ... up to the last date on 24th September.

A forecast can be produced on any of these dates for lead times up to 40 days in
advance, but in this study, only lead times up to 4 weeks are considered. This is
because such a time period is relatively reliable and adequate for use by farmers
especially in Nigeria to plan and make decisions about farming practices such as
planting, weeding, fertilizer application, etc. It should be noted also that at longer
lead times, forecasts tend to lose accuracy (Inness and Dorling 2010), and so, the
weekly to one month lead time forecast chosen is likely to be more reliable than lead
time more than one month. In this study, forecast made with a lead time of one week
(referred to as lead1) means the forecast is produced on a particular week and is valid
for one week (up to the following week). It is also possible to produce a lead 1, 2, 3,
or 4 forecast on any of the 20 start dates from May 14 to September 24. The choice of
any of the start dates depends on the period of interest within the season and the goal
which the user or client wants to achieve or satisfy.

The correlation between the observed and forecast anomalies (ACC) at different
lead times and with persistence forecast is shown in Fig. 4.

Seasonal mean ACC values tend to be generally poor and less than 0.5 but weekly
values tend to be better in some weeks at different lead times. For example, start dates
of week 8 (July 2), week 12 (July 30), week 13 (August 6), week 14 (August 13), and
week15 (August 20) for lead 1 forecast have the highest and best ACC values of 0.71,
0.84, 0.75, 0.72, and 0.65, respectively, which suggest a strong positive relationship
between forecast and observation. These values are all significant at 0.01 level while
the ACC value of 0.54 in week 11 (July 23) is significant at 0.05 level. High ACC
scores were also observed for a few start dates at lead times 2, 3, and 4. The scores
generally decreased with increasing lead time (Fig. 4) and the poorest scores were

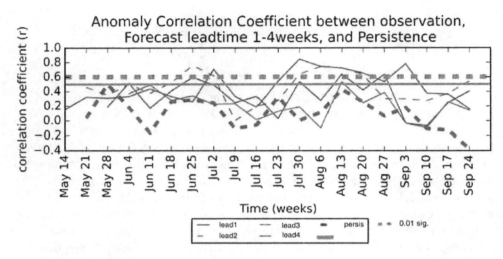

Fig. 4 Anomaly Correlation between forecast at different lead times compared with persistence forecast. Topmost thick horizontal grey dotted line indicates ACC values that are statistically significant at 0.01 while the solid grey line represents values that are significant at 0.05 level (Source: Ugbah 2016)

observed in lead 4 forecasts with a few negative correlations. This is consistent with Li and Robertson (2015). A statistical test to determine the level of significance of these values was done using the Statistical Package for Social Sciences (SPSS) software.

Persistence is also a baseline, like mean climatology, used to compare with forecasts in order to determine their quality. The persistence forecast for lead 1 was produced by persisting observation of the present week for the following week's forecast. The result was then correlated with the forecast at different lead times. It is often said that the long time range involved in subseasonal forecasting has made it difficult for it to beat Persistence forecast (Vitart 2014) but result from this study has proven otherwise as the forecast at the different lead times beat persistence in most of the weeks This approach according to Jolliffe and Stephenson (2003) only performs well for short range forecast, but result from this work support the view that it can also perform well for medium ranged forecast as shown (Fig. 4).

Comparison Between Observed and Forecast during El Nino and La Nina Years

ENSO and MJO both interact in modulating subseasonal to seasonal precipitation. It is assumed that since the ECMWF model is known for its high skill in simulating MJO events, it is also likely to accurately simulate subseasonal rainfall variability during different phases of ENSO. Based on this, it will be good to look at how the model forecast for precipitation during El Nino and La Nina years either agree or differ from the lead 1 forecast for the 1995–2015 period. Only lead one forecast is chosen because it shows the least bias with observed climatology compared to lead times of two, three, and four weeks. El Nino and La Nina years were identified using the Oceanic Nino Index (ONI) as earlier discussed. Four La-Nina (1998, 1999, 2000, 2010) and Five

El-Nino events (1997, 2002, 2004, 2009, 2015) were identified and the forecast for these years extracted and plotted for comparison. The selection of 1997 El Nino and 1999 La Nina events is in agreement with Enso events between 1963–1999 identified by Joly and Voildoire (2009). The strength of ENSO usually weakens towards the end of first quarter of the year and reaches maximum strength around December to February (Joly and Voildoire 2009). A strong African monsoon is linked to La Nina events which are observed in April, May, or June (Joly and Voildoire 2009).

The result of analysis shows that both La Nina and El Nino years forecast are in reasonable agreement with observed precipitation during the first eight weeks (May14–July2) and show a wetter bias, but begin to show marked differences beyond this period particularly in mid-July and in August (Fig. 5) when the little dry season is usually experienced in the study area.

The differences between them were calculated as bias and compared with the 1995–2015 climatology shown in Table 2.

The seasonal mean biases were all negative suggesting underestimation of precipitation by the model compared to observation and was higher (−0.97) during La Nina years than El Nino (−0.41). The 1995–2015 period as a whole, however, showed lower bias compared to the La Nina years. It is likely that the larger bias might be by chance rather than a certainty. The El Nino years showed better results with the least bias and RMSE. The seasonal mean RMSE in La Nina years was 2.24 compared to El Nino which has 1.84 (Table 2).

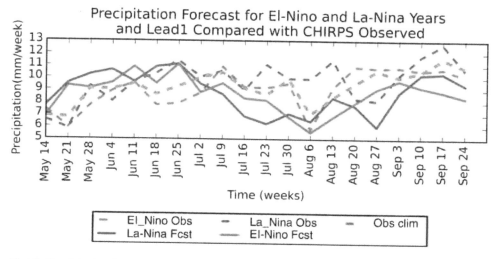

Fig. 5 Precipitation forecast and observation for El Nino and La Nina years for one week lead time compared with observed climatology (Source: Ugbah 2016)

Table 2 Comparing Lead1 scores for the climatological period, El Nino, and La Nina years

Scores of the ECMWF subseasonal weekly forecast over Southern Nigeria			
Period	1995–2015	El Nino years	La Nina years
BIAS	−0.69	−0.41	−0.97
RMSE	1.63	1.84	2.24

Weekly biases compared were positive during the first 7–8 weeks after which they became negative. This is to say that forecast for both El Nino and La Nina years was over predicted during the initial weeks and under-predicted thereafter. Another observation is that the forecast seem to follow the pattern of climatology rather than observation and was lower than the observed for a greater part of the season. The high bias and RMSE particularly in La Nina years shows that the model might be worst in La Nina years and more reliable in El Nino years as the El Nino forecast tends to agree more with the observation and climatology than forecast during La Nina years. All these still point to the model's inability to accurately simulate southern Nigeria precipitation and suggest that there is still a missing tele-connection between precipitation in the study area and ENSO that has not yet been properly resolved by the ECMWF model. This, however, does not imply that the model is bad, but requires fine-tuning to correct some of the biases observed in order to improve the quality of the forecast. Such biases could arise from the poor represen-tation of the ENSO, MJO, and precipitation tele-connection within the study area. Another reason could be incorrect parameterization of convection and processes which modulate weather over Nigeria. This can also adjusted to give better results. The timing of ENSO events and the delay in the response of a particular region to an ENSO phase is also an important consideration in simulating the tele-connection between them (Joly and Voildoire 2009). This is because significant tele-connection is sometimes observed during the developing stage of ENSO or during its terminal stage (Joly and Voildoire 2009). All these, still remain a big challenge to many numerical modeling experts trying to simulate tele-connection between ENSO and the West African Monsoon (Joly and Voildoire 2009). The "August break" is well captured by the model forecast with strong agreement during the El Nino years than La Nina. However, the greatest noise is observed during this period (July–August). This again shows that the model can be improved.

Performance of the S2S Model Precipitation Forecast During Rainfall Peaks and Period of Little Dry Season

Should the ECMWF S2S model be able to properly simulate high rainfall amounts and pattern during periods of maximum rainfall within the season, then it will no doubt be a good tool in providing useful information for extreme rainfall disaster preparedness, prevention, and risk reduction. Fortunately, the model is able to properly simulate pattern of rainfall in the region and also identify periods of peak rainfall which mostly occurs in June and September as shown in Figs. 3, 4, and 5. The model is also able to depict the little dry season period which occurs between July and August. Such information is considered useful to planners. But unfortu-nately the model shows a low skill in simulating rainfall amount especially during the peak rainfall period and the little dry season. This is evident in the large biases observed between last week in June and mid-August at lead time of two, three, and four weeks. The biases and RMSE are however minimal at lead time of one week which is good for early warning purposes and disaster preparedness at least one week before the event happens. The skill is also found to be poor during La Nina years which are notable for high rainfall than El-Nino years in the region. This means that the model is not very skillful in simulating heavy precipitation amounts during heavy rainfall periods especially at lead times beyond one week. It is however still possible that some adjustment in the model by ECMWF to correct the observed biases could give better and reliable results which would make that ECMWF S2S forecast model a dependable tool for predicting extreme high/low rainfall and thus helpful in adapting and mitigating the effects of climate change in the study area.

Conclusion

Large negative biases and root mean square error which were observed to increase with lead times show that the model generally has a dry bias towards estimating precipitation over the area. The high biases and errors observed during La-Nina, than El-Nino years suggests that the model performs better during El-Nino years. Users should therefore be cautious of its use during La-Nina years. Re-calibrating the model output to reduce these biases is likely to produce better results. Seasonal mean Bias and RMSE which generally increased with increasing lead time during the season showed the best results in the one week lead time and the worst in four weeks lead time. This suggests that the forecast was better at shorter lead time. Seasonal mean ACC values which were generally less than 0.5 but with weekly values greater than 0.7 on certain start dates of the lead 1 and 2 forecast, and lower values for leads 3 and 4 confirm that the forecast is best at lead time of one and two weeks during specific periods in the season than at longer lead times. ECMWF should also consider making some adjustments in the model in aspects of parametization of ENSO tele-connection with rainfall over the study area to reduce the large biases observed during La Nina years. This is important because La Nina events bring more rains to the region than El Nino events and so the opportunity of more rains and water availability during La Nina years should be maximally exploited through use of good and reliable prediction tools that will provide accurate and reliable information needed for planning by farmers, environmentalist, disaster managers, water resource managers, and policy makers. The results also show that the S2S model precipitation forecast has a skill which is higher than persistence forecast at all the lead times of one to four weeks, thus negating the statement by Vitart (2014) that the long time range involved in subs-seasonal fore-casting has made it difficult for it to beat Persistence forecast.

The observed large negative biases and RMSE at lead times of 2, 3, and 4 weeks suggest that the model is not yet very skillful in simulating precipitation amounts over southern Nigeria at longer time ranges; it is, however, very skillful in simulation precipitation amount, pattern, and periods of rainfall peaks/minima during the season at lead time of 1 week. This is evident in the very low biases and errors, thus confirming that the best results were observed for precipitation forecast made with a lead time of one week only. This means that S2S precipitation forecast can provide one week advance signals of periods of peak rainfall, little dry season and expected rainfall amount over southern Nigeriais. Such information is considered to be a good early warning and disaster preparedness tool. It could also be a useful tool for planning and decision making if provided at least one week before the event happens. Thus, a way of mitigating and adapting to the effects of climate variability and change is used. The ECMWF S2S can therefore be recommended for this purpose and for updating seasonal forecast on weekly basis in Nigeria for maximum Agricultural production and for disaster reduction since the skill is high at lead time of one week. It is therefore a good tool for monitoring of rainfall during the rainy season and could still yield better results at longer lead time if the model is further improved by ECMWF.

Future work will expand the scope to cover the entire Nigeria and also assess the skill of the individual ensemble members of the model instead of the average of the ensembles used in this study. Rain guage observation data will be considered for comparison with forecast. It is also good for work in the future to include determination of rainfall onset, cessation, and length of season using the S2S data instead of looking only at how the model can simulate rainfall amount and pattern as done in this study.

References

Ayanlade A, Adeoye NO, Babatimehin O (2013) Intra-annual climate variability and malaria transmission in Nigeria. Bull of Geogr Socio-economic series 21(21):7–19
Chai T, Draxler RR (2014) Root mean square error (RMSE) or mean absolute error (MAE)? Arguments against avoiding RMSE in the literature. Geosci Model Dev 7:1247–1250
Chidiezie T, Shrikant C (2010) West African monsoon: is the August break "breaking" in the eastern humid zone of Southern Nigeria? Climate Change 103:555–570
Eludoyin OM, Adelekan IO (2013) The physiologic climate of Nigeria. Int J Biometeorol 57:241–264. Springer
Funk CC, Peterson PJ, Landsfeld MF, Pedreros DH, Verdin JP, Rowland JD, Romero BE, Husak GJ Michaelsen JC, Verdin AP (2014) A quasi-global precipitation time series for drought monitoring. US Geol Surv Data Ser 832:4
Haiden T, Janousek M, Bauer P, Bidlot J, Ferranti L, Hewson T, Prates F, Richardson DS, Vitart F (2015) Evaluation of ECMWF forecasts, including 2014–2015 upgrades ECMWF Technical Memoranda, No. 765. Available at http://www.ecmwf.int/sites/default/files/elibrary/2015/15275-evaluation-ecmwf-forecasts-including-2014-2015-upgrades.pdf]. Reading
Hamill TM (2012) Verification of the TIGGE multimodel and ECMWF reforecast-caliberated probabilistic precipitation forecast over Contiguos United States. Month Weath Rev 140:2232–2252
Hansen JW (2002) Realizing the potential benefits of climate prediction to agriculture: issues, approaches. Agricultural Systems 74:309–330
Inness P, Dorling S (2010) Operational Weather Forecasting, Wiley-Blackwell, 231pp
Intergovernmental Panel on Climate Change (IPCC) 2018 Summary for policymakers. In: Masson-Delmotte V, Zhai P, Pörtner HO, Roberts D, Skea J, Shukla PR, Pirani A, Moufouma-Okia W, Péan C, Pidcock R, Connors S, Matthews JBR, Chen Y, Zhou X, Gomis MI, Lonnoy E, Maycock T, Tignor M, Waterfield T (eds) Global warming of 1.5°C. An IPCC Special Report on the impacts of global warming of 1.5°C above pre-industrial levels and related global greenhouse gas emission pathways, in the context of strengthening the global response to the threat of climate change, sustainable development, and efforts to eradicate poverty. In Press
International Research Institute (IRI) (2016) CHIRPS daily precipitation data. Available at http://iridl.ldeo.columbia.edu/SOURCES/.UCSB/.CHIRPS/.v2p0/.daily/global/.0p05/index.html?Set-Language=en
Jolliffe IT, Stephenson DB (2003) Forecast verification – a practitioner's guide in atmospheric science. Wiley and sons, Chichester, 247pp
Joly M, Voildoire A (2009) Influence of ENSO on the west African monsoon: temporal aspects and atmospheric processes. J Clim 22:3193–3210
Li S, Robertson AW (2015) Evolution of submonthly precipitation forecast skill from global ensemble prediction system. Month Weath Rev 143:2871–2889

Maue RN, Langland RH (2014) Northern hemisphere forecast skill during extreme winter weather regimes – a presentation made at the 94th American Meteorological Society annual meeting, Atlanta. Available at http://www.models.weatherbell.com/news/maue_AMS_2014.ppt

Molteni F, Vitart F, Lang S, Weisheimer A, Keeley S (2016) Sub-seasonal prediction at ECMWF: Present, Past (recent and less recent). Accessed June 2016 [Available at http://www.ecmwf.int/sites/default/files/elibrary/2015/14495-sub-seasonal-prediction-ecmwf-present-past-recent-and-less-recent-and-future.pdf]

NOAA (2016) National Oceanic and Atmospheric Administration Oceanic Nino Index data [available at: https://origin.cpc.ncep.noaa.gov/products/analysis_monitoring/ensostuff/ONI_v5.php]

NiMet (2013) Nigerian meteorological Agency: Seasonal Rainfall Prediction Brochure. Available at http://nimet.gov.ng/sites/default/files/publications/2013-seasonal-rainfall-prediction.pdf. Abuja

NiMet (2014) Nigerian meteorological Agency: Seasonal Rainfall Prediction Brochure. Available at http://www.nimet.gov.ng/sites/default/files/publications/SRP%20BOCHURE%20FINAL.pdf. Abuja

Siegmund J et al (2015). Toward a seasonal precipitation prediction system for West Africa: Performance of CFSv2 and high-resolution dynamical downscaling. J Geoph Res Atm 120 (15):7316–7339. AGU Publications

Takaya Y (2015) The sub-seasonal to seasonal (S2S) prediction project. Presentation at World Meteorological Organization workshop held at Pune, India, 9–11 November 2015. Available at http://www.wmo.int/pages/prog/wcp/wcasp/documents/workshop/pune2015PPT/day1/Session2-Takaya_WMO_workshop_S2S_201511.Pdf

Ugbah PA (2016) Dissertation: evaluation of European Centre for Medium-range Weather Forecast (ECMWF) subseasonal to seasonal forecast over Nigeria, Department of Meteoorology, University of Reading, 44

Vitart F (2004) Monthly forecasting at ECMWF. Month Weather Rev 132(12):2761–2779

Vitart F (2005) Monthly forecast and the summer 2003 heat wave over Europe: a case study. Atm Sci Lett 6:112–117. Wiley publishers

Vitart F (2014) Evolution of ECMWF sub-seasonal forecast skill scores. Q J Meteorol Soc 140:1889–1899. Royal Met Society

Vitart F, Ardilouze C, Bonet A, Brookshaw A, Chen M, Codorean C, Deque M, Ferranti L, Fucile E, Fuentes M, Hendon H, Hodgson J, Kang H, Kumar A, Lin H, Liu G, Liu X, Malguzzi P, Mallas I, Manoussakis M, Mastrangelo D, MacLachlan C, McLean P, Minami A, Mladek R, Nakazawa T, Najm S, Nie Y, Rixen M, Robertson A, Ruti P, Sun C, Takaya Y, Tolstykh M, Venuti F, Waliser D, Woolnough S, Wu T, Won D, Xiao H, Zaripov R, Zhang L (2016) The Sub-seasonal to Seasonal Prediction (S2S) Project Database. Bull Amer Meteor Soc https://doi.org/10.1175/BAMS-D-16-0017.1, in press

12

Plants and Plant Products in Local Markets within Benin City and Environs

Moses Edwin Osawaru and Matthew Chidozie Ogwu

Contents

Abstract

The vulnerability of agriculture systems in Africa to climate change is directly and indirectly affecting the availability and diversity of plants and plant products available in local markets. In this chapter, markets in Benin City and environs were assessed to document the availability of plants and plant products. Markets were grouped into urban, suburban, and rural with each group having four markets. Majority of the plant and plant product vendors were women and 88 plant species belonging to 42 families were found. Their scientific and common names were documented as well as the parts of the plant and associated products available in the markets. Most of the plant and plant products found in local

M. E. Osawaru
Department of Plant Biology and Biotechnology, Faculty of Life Sciences, University of Benin, Benin City, Edo State, Nigeria

M. C. Ogwu (✉)
Department of Plant Biology and Biotechnology, Faculty of Life Sciences, University of Benin, Benin City, Edo State, Nigeria

Scuola di Bioscienze e Medicina Veterinaria, Università di Camerino – Centro Ricerche Floristiche dell'Appennino, Parco Nazionale del Gran Sasso e Monti della Laga, Barisciano (L'Aquila), Italy
e-mail: matthew.ogwu@uniben.edu

markets belong to major plant families. Urban markets had the highest diversity of plants and plant products. Three categories of plants and plant products were documented. Around 67% of the plants and plant products were categorized as whole plant/plant parts, 28% as processed plant parts, while 5% as reprocessed plant/plant parts. It was revealed that 86% of these plants are used as foods, 11% are for medicinal purposes, while 3% is used for other purposes. About 35% of plants and plant products across the markets were fruits, which is an indication that city and environs are a rich source of fruits. The local knowledge and practices associated with the plants and plant products can contribute towards formulating a strategic response for climate change impacts on agriculture, gender, poverty, food security, and plant diversity.

Keywords

Climate change · Ethnobotany · Plant diversity · Plant products · Food security · Market survey · Indigenous plants species · Economic plants · Agriculture vulnerability · Sustainable development

Introduction

The utilitarian nature of humans is driving the massive extinction of biodiversity, climate change, and ecological vulnerability. However, a greater understanding of plant-human interactions can contribute to sustainable development, addressing climate change and biodiversity loss, food security, and poverty reduction. All plants are considered important and can potentially serve to fulfill one or more of our basic needs – food, shelter, and clothing as well as environmental integrity. Plant product refers to goods and services derivable from plants and may include whole plant or plant part (used as ingredients and condiments). Proper local and scientific identification of plant materials is necessary to determine and predict the role of a plant and this will require a general knowledge of botany, sociology, and anthropology. Plant is essential for our continued survival on earth as they directly or indirectly provide food for survival, medicine, fibers, chemical products, and other commodities as well as to protect and maintain the environment against erosion, used to cure disease and relieve from suffering. Many industries are dependent on plants for their raw materials. Some of the most outstanding materials of modern civilization are obtained from plants, such as wood, tanning materials and dyestuffs, oils, resins, gums, varnishes, beverages, etc. Plants provide raw material for industrialization and are basis of the green revolution and a pillar for food security. The esthetic value of plants has no small influence on man's overall life satisfaction, as evidenced by the host of garden enthusiasts and flower lovers. Plants are also the basis of a vegan lifestyle.

In the economy of nature, the production and distribution of plant products have a profound influence on the environmental, economic, and social life of a nation with both domestic and international influence. The maintenance of an adequate supply of food and plant-based industrial raw materials is essential to the existence, as well as the prosperity, of any nation (Burkill 1985). Additionally, plants also have important roles in the tribal, social and cultural life of man (Osawaru and Dania-Ogbe 2010; Osawaru and Ogwu 2014a). Local markets are an integral part of life and cultural practice of the people especially in developing countries as a social, economic, and ecological institution. Plant products available in the market can be used as an indicator for biodiversity richness, climate changes effects, and agricultural vulnerabilities. This is more important in Africa, which according to Ogwu (2019) the environment and agriculture systems are most vulnerable climate change. Markets in rural parts of Africa are often scheduled at considerable day interval, whereas in urban and semi-urban centers, it is mostly open every day or night. Sellers have their stalls or place while hawkers also patrol the market with their various plants and plant products. Markets are rich sources of information on plants and plant products as well as an easily accessible and cost-effective place for plant-based fieldwork and germplasm collection. Markets can provide qualitative and quantitative data concerning cultural, social, and economic aspects of a plant's usage (Bye and Linares 1983; Martin 1992; Cunningham 2001). Moreover, markets are recognized as a vital botanical record of the history of useful plants in a region (Whitaker and Cutler 1966). They are places of intensive interaction between people and plants.

Local traders (mostly women) are very knowledgeable about the uses of plants and their seasonal availability. This knowledge is vital in the global response to global climate change and massive loss of biodiversity in this sixth extinction era. In Nigeria and West Africa, market vendors are known to deal in certain types of plants and plant product and are found clustered together, which imposes a sort of market influence such as fixed prices for their commodities. However, traders have certain concession such as monopoly to regular customer and slight price variations. Moreover, the survey of marketplaces provides information about food and nutritional value of plant and products as well as their ethnobotany (Nguyen 2005). Findings from such studies have been used to draw interesting conclusions and hypothesize about human-environmental-plant interactions and relationships. Climate change, migration, and economic forces can influence the availability of certain plant and plant products in the market. Obiri and Addai (2007) surveyed economic plants in Kumasi central market and documented a total of 150 plant species from 55 families most of them had multiple uses – 57% and 20% used for medicinal and food purposes, respectively. Idu et al. (2010) documented the medicinal plants sold in markets in Abeokuta, Nigeria, revealed 60 medicinal plant species used for traditional health management. The ethnobotanical survey of Yaradua and El-Ghani (2015) reported 54 plants belonging to 33 families from Katsina metropolis markets. The objective of this chapter is to identify and document the plants and plant products sold in local markets in Benin City and environs.

This chapter will compare the diversity of the plants and plant products available in local markets in urban, peri-urban, and rural centers in Benin City, Southern Nigeria. Thereafter, this chapter will categorize the plants and plant products based on the level of processing it was subjected to as well as their taxonomic families. The results will seek to promote local markets as a reservoir of plant germplasm and contribute towards understanding of how climate change vulnerability is affecting agriculture system in Edo state, diversity of plant and plant-based food materials available in local markets that can potentially contribute to addressing food security, poverty, and sustainable development. It will also highlight the plant parts and plant products sold in the markets.

Plants and Plant Products in Local Markets: The Case of Benin City and Environs

There are numerous open markets in Benin City (latitude 06° 19′00″ E to 6° 21′00″ E and longitude 5° 34′00″ E to 5° 44′00″ E; average elevation of 77.8 m above sea level; 2006 est. pop. 1,147,188 with an annual growth rate of 2.9%), which is one of the oldest cities in Nigeria and the capital of Edo State, Southern Nigeria. It is within the tropical rainforest zone of Nigeria with an estimated area of 550 km^2. Geologically the city has a sedimentary formation of the Miocene-Pleistocene age. Benin has an undulating topography with a vegetation type characterized by lowland rainforest and an annual average rainfall of 800 mm. A 35-year study by Floyd et al. (2016) revealed massive climate fluctuations especially in average rainfall, temperate, humidity and suggested that it is impacting plant production and environmental changes due to soil erosion. The fluctuation is attributed to high anthropogenic activities in Benin City (Efe and Eyefia 2014). With increased warming, flooding, and urbanization, agriculture and food production are threatened in Benin City (Atedhor et al. 2011).

Benin is the center of Nigeria's rubber industry, but processing palm nuts for oil is also an important traditional industry. Benin has numerous local markets strewn across the city to cater to the needs of its inhabitants as well as to serve as a sales outlet for the numerous farm produce cultivated in the rural areas of the state as well as in urban home gardens. The nodal nature of Benin makes it an ideal place for various commercial activities as these farm produce can be easily transported to cities like Lagos, Abuja, and Port Harcourt. Benin City is endowed with a wide diversity of plants and plant products. The sales of plants and plant products play a key role in the sustenance of livelihoods of people providing income, employment, food, and medicines among others in Benin City and environs. Some of the local markets in Benin City and environs are God's Market (Ekiosa), Oba Market (Ekioba), New Benin Market, Santana Market, Uselu Market (Ediaken Market), Oliha Market, Ugbogiobo Market, Evbuotubu Market, Oregbeni Market (Ikpoba Hill Market), Ekiadolor Market, Iguobazuwa Market, Ehor Market, and Usen Market.

The sampling frame considered markets within Benin City and environs, which were delimited into three categories according to the status defined by Osawaru and Odin (2012) – urban, peri-urban, and rural. A reconnaissance visit was undertaken to all the local markets in Benin City and environs. Twelve markets were randomly selected for sampling. They consist of four urban, four peri-urban, and four rural markets (Table 1).

Table 1 Sampling sites for the survey of plants and plant products in Benin City and environs

Market	Category	Local government area
New Benin	Urban	Oredo
Uselu	Urban	Egor
Oba	Urban	Oredo
Oregbene	Urban	Ikpoba-Okha
Evbuotubu	Peri-urban	Egor
Ugbogiobo	Peri-urban	Ovia North East
Ugbiohioko	Peri urban	Egor
Iguobazuwa	Peri-urban	Ovia South West
Ehor	Rural	Uhunnwode
Usen	Rural	Ovia North East
Ekiadolor	Rural	Ovia North East
Ugbogui	Rural	Ovia South West

In each market, ten traders of mixed age and sex were randomly selected and plants and plant products in their stalls were assessed. Each market was visited three times. First, to map out the randomly selected informant, secondly, to administer the questionnaire and inventory the plants and plant products, and finally, to seek clarity for some questions outlined in the questionnaire. Responses via the questionnaires were retrieved from the questionnaires, translated and scored by typing into Microsoft Excel, and analyzed quantitatively. Plants and plant products were categorized according to Osawaru and Odin (2012) i.e.,

1. 1^0 of plants and plant products-whole plant/plant part
2. 2^0 of plant and plant product-processed plant part
3. 3^0 of plant and plant product-reprocessed plant/plant part

Majority of the traders encountered in the markets were women. This confirms the findings of De Caluwe (2011) and Agea et al. (2011) that trading in plant and plant products are dominated by women. In a different study, Osawaru and Ogwu (2014b) also established that women contribute significantly to holding plant germplasm. These findings underscore the importance of women in the fight to address the effects of climate change especially food security and sustainable agriculture. Moreover, the sales of plants and plant products in the different markets were practiced by different tribes and ethnic groups in all the markets surveyed.

A total of 88 plants and plants product was found in all the local markets assessed (Table 2). The botanical and common names, forms, diversity, and categories of the plant and plant products sold in local markets within Benin City and environs are presented in Table 2. These 88 plants and plant product are distributed into 42 families. Presence of plant and plant products varies in the different markets. For instance, *Adansonia digitata, Brassica oleracea, Cucumis sativus, Cucurbita pepo, Cyperus esculetus, Dialium guineense, Ricinus communis, Pennisetum glaucum, Pentaclathra macrophylla, Myristica fragans,* and *Phoenix dactylifera*

Table 2 Diversity of plants and plant products in 12 local markets within Benin City and environs

Botanical name	Family	Common name	Form/product type	Local name (Bini)	Urban market	
					New Benin	Uselu
Abelmoshus esculentus L.	Malvaceae	Okra		Ikhiav-bo	+	+
Adensonia digitata L.	Malvaceae	Baobab			+	+
Allium cepa L.	Liliaceae	Onion		Alubara	+	+
Allium sativum L.	Liliaceae	Garlic			+	+
Amaranthus caudatus L.	Amaranthaceae	Spinach		Ebaafor	+	+
Anacardium occidentalis L.	Anacadiaceae	Cashew			+	−
Ananas comosus L.	Bromeliaceae	Pineapple		Edinebo	+	+
Annona muricata L.	Annonaceae	Soursop			+	+
Arachis hypogaea L.	Fabaceae	Groundnut		Isaerewe	+	+
Azadirachta indica L.	Meliaceae	Neem			−	+
Bombax buonopozense L.	Malvaceae				−	−
Brassica oleracea L.	Brassicaceae	Cabbage			+	+
Calotropis.procera Auton.	Apocynaceae				+	+
Capsicum annum L.	Solanaceae	Pepper		Ehien	+	+
Capsicum frutescens L.	Solanaceae	Pepper		Ikpovb-ukho	+	+
Carica papaya L.	Caricaeae	Pawpaw		Uhoro	+	+
Celosia argentea L.	Amaranthaceae	Celosia			+	+
Citrus aurantifolia L.	Rutaceae	Lime		Alimonegiere	+	+
Citrus limon L.	Rutaceae	Lemon			+	+
Citrus sinensis Osbeck	Rutaceae	Orange		Alimebo	+	+
Cochorus olitorius L.	Tiliaceae	Jute			+	+
Cocos nucifera L.	Palmae	Coconut		Ivin	+	+
Cola acuminate Engl.	Sterculiaceae	Kolanuts		Gbanja	+	+
Cola nitida Schum.	Sterculiaceae	Kolanut		Evbedo	+	+
Colocasia esculenta Schott	Araceae	Cocoyam		Akaha	+	+
Crescentia cujele L.	Curcubitaceae	Calabash		Uko	+	+
Cucumeropsis mannii Naudin	Cucurbitaceae	Melon		Ogi	+	+
Cucumis sativus L.	Cucurbitaceae	Cucumber			+	+
Cucurbita pepo L.	Cucurbitaceae				+	+
Cucurma longa L.	Zingiberaceae				+	+
Cymbopogon citrates L.	Poaceae	Lemmon grass		Ebiti	+	+
Cyperus esculentus L.	Cyperaceae	Tiger nut			+	+
Dacryodes edulis Lam	Burderaceae	African pear		Oruvbu	−	−
Daucus carota L.	Apiaceae	Carrot			+	+
Dennettia tripetala Bak. F.	Annonaceae	Pepper fruit		Ako	+	+
Dialium guineense Willd	Fabaceae	Velvet tamarind			+	+
Dioscorea alata Lour	Dioscoreaceae	Water yam		Igierua	+	+
Dioscorea cayenensis Lam.	Dioscoreaceae	Aerial yam		Ikpen	+	+

		Peri-urban market				Rural market			
Oba	Oregbene	Evbuotubu	Ugbogiobo	Ugbiohioko	Iguobazuwa	Ehor	Usen	Ekiadolor	Ugbogui
+	+	+	+	+	+	+	+	+	+
+	+	+	−	−	−	−	+	+	−
+	+	+	+	+	+	+	+	+	+
+	+	+	+	+	+	+	+	+	+
+	+	+	+	+	+	+	+	+	+
+	−	+	+	−	+	+	+	+	+
+	+	+	+	+	+	+	+	+	+
+	+	+	−	−	−	−	−	+	+
+	+	+	+	+	+	+	+	+	+
+	−	+	+	+	+	+	−	−	−
−	−	+	+	+	+	−	+	+	+
+	+	+	−	+	−	−	−	−	−
+	−	−	−	−	+	−	+	−	−
+	+	+	+	+	+	+	+	+	+
+	+	+	+	−	−	−	−	−	−
+	+	+	+	+	+	+	+	+	+
+	+	+	+	+	+	+	+	+	+
+	+	+	+	+	+	+	+	+	+
+	+	+	+	−	−	−	+	+	−
+	+	+	+	+	+	+	+	+	+
+	+	+	+	+	+	+	+	+	+
+	+	+	+	+	+	+	+	+	+
+	+	+	+	+	+	+	+	+	+
+	+	+	+	+	+	+	+	+	+
+	+	+	+	+	+	+	+	+	+
+	+	−	+	+	+	+	−	−	−
+	+	+	+	+	+	+	+	+	+
+	+	+	+	−	−	−	−	+	−
+	+	+	+	−	−	+	−	−	−
+	+	−	−	+	−	−	−	−	−
+	+	+	−	−	−	+	−	−	+
+	+	−	−	−	−	−	−	−	−
−	−	−	+	+	+	+	+	+	−
+	+	+	+	+	+	−	−	+	−
−	+	−	+	+	−	+	+	−	+
+	+	−	+	+	+	−	−	−	−
+	+	+	+	+	+	+	+	+	+
+	+	+	+	+	+	+	+	+	+

(continued)

Table 2 (continued)

Botanical name	Family	Common name	Form/product type	Local name (Bini)	Urban market	
					New Benin	Uselu
Dioscorea rotundata Poir	Dioscoreaceae	Yam	Yam chips and yam flour	Emowe	+	+
Elaeis guineensis Jacq	Palmae	Oil palm	Oil	Udin	+	+
Garcina cola Heckel	Guittiferae	Bitter cola		Edun	+	+
Glycine max L.	Fabaceae	Soy bean		Owerie-otan	+	+
Gnetum africanum Welw.	Gnetaceae				+	+
Gossypium hirsutum L.	Malvaceae	Cotton		Oruhu	+	+
Hibiscus cannabinus L.	Malvaceae				+	+
Hibiscus sabdarifa L.	Malvaceae	Roselle		Zobo	+	+
Ipomea batata L.	Convolvulaceae	Sweet potato		Iyinebo	+	+
Irvingia gabonensis Baill	Irvingiaceae	Bush mango		Ogwi	+	+
Lycopersicum esculentum L.	Solanaceae	Tomato	Tomato paste	Etomat-osi	+	+
Mangifera indica L.	Anacadiaceae	Mango		Emango	+	+
Manihot esculatua Crantz	Euphobiaceae	Cassava	Garri, fufu, bobozi and cassava flour	Igari	+	+
Murraya koenigii L.	Rutaceae	Curry		Curry leaf	+	+
Musa paradisiaca L.	Musaceae	Banana			+	+
Musa sapientum Linn	Musaceae	Plantain	Chips and plantain flour	Oghede	+	+
Myristica fragrans Houtt.	Myristicaceae	Nutmeg			+	+
Ocimum gratissimum Linn	Lasiottae	Scent leaf		Ebihiri	+	+
Oryza sativa L.	Poaceae	Rice		Izee	+	+
Parkia clappertoniana Keay	Fabaceae	Locust bean		Evbarie	+	+
Pennisetum glaucum (L.) R. Br.	Poaceae	Millet	Kunu		+	+
Pentaclethra macrophylla L.	Fabaceae	African oil bean			+	+
Persea americana Miller	Lauraceae	Avocado Pear			+	+
Phaseolus vulgaris L.	Fabaceae	Beans	Beans cake, beans flour	Ere	+	+
Phoenix dactylifera L.	Palmae	Date palm			+	+
Piper guineense Schumach	Piperaceae	African pepper		Oziza	+	+
Psidium guajava L.	Myrtaceae	Guava			−	−
Rauwolfia vomitoria Afzel.	Apocynaceae	Rauwolfia		Akata	−	+
Ricinus communis L.	Euphorbiaceae	Castor oil			+	+
Saccharum officinarum L.	Solanaceae	Sugar cane	Sugar	Ukhure	+	+
Sesanum orientale L.	Pedaliacaece				+	+
Solanum melogena L.	Convolvulaceae	Garden-egg		Ekhue	+	+
Solanum tuberosum L.	Poaceae	Irish potato			+	+
Sorghum bicolour L.	Poaceae	Guinea corn	Kunu		+	+
Spondias mombin L.	Meliaceae	Hug phem		Okhikhan	−	+

		Peri-urban market				Rural market			
Oba	Oregbene	Evbuotubu	Ugbogiobo	Ugbiohioko	Iguobazuwa	Ehor	Usen	Ekiadolor	Ugbogui
+	+	+	+	+	+	+	+	+	+
+	+	+	+	+	+	+	+	+	+
+	+	+	+	−	+	−	+	−	+
+	+	+	+	+	+	+	+	+	+
+	+	+	−	−	+	+	+	−	+
+	+	+	+	+	+	+	+	−	+
+	−	−	−	−	−	+	−	+	−
+	+	+	+	+	−	−	+	−	+
+	+	+	+	+	+	+	+	+	+
+	+	+	+	+	+	+	+	+	+
+	+	+	+	+	+	+	+	+	+
+	+	+	+	+	+	+	+	+	+
+	+	+	+	+	+	+	+	+	+
+	+	+	+	+	+	+	+	+	+
+	+	+	+	+	+	+	+	+	+
+	+	+	+	+	+	+	+	+	+
+	+	+	−	+	−	−	−	−	−
+	+	+	+	+	+	+	+	+	+
+	+	+	+	+	+	+	+	+	+
+	+	+	+	+	+	+	+	+	+
+	+	−	−	−	−	−	−	−	−
+	+	−	−	−	−	−	−	−	−
+	+	−	+	+	+	−	−	−	+
+	+	+	+	+	+	+	+	+	+
+	−	−	−	−	−	−	−	−	−
+	+	−	−	−	+	+	−	+	−
+	+	+	+	+	−	−	−	−	−
+	+	−	−	−	−	+	+	+	+
+	−	−	−	−	−	−	−	−	−
+	+	+	−	+	+	−	−	−	+
+	+	+	+	+	−	−	−	−	−
+	+	+	+	+	+	+	+	+	+
+	+	+	−	+	−	−	−	−	−
+	−	−	−	−	−	−	−	−	−
+	+	+	+	+	−	−	−	−	−

(continued)

Table 2 (continued)

Botanical name	Family	Common name	Form/product type	Local name (Bini)	Urban market	
					New Benin	Uselu
Talfairia occidentalis Hook	Cucurbitaceae	Pumpkin		Uvbeg-hen	+	+
Talinum triangulare Jacq	Portulaceae	Water leaf		Ebodo-don	+	+
Tamarindus indica L.	Fabeaceae				+	+
Tetrochidium didymostemon (Baill.) Pax & K. Hoffm	Euphorbiaceae				–	–
Thaumatococcus danielli Benth.	Marantaceae			Ebe-eba	+	+
Theobroma cacao L.	Sterculiaceae	Cocoa		Koko	+	–
Thymus vulgaris L.	Lamiaceae	Thyme			+	+
Treculia Africana Decne.	Moraceae	African breadfruit			+	+
Trilepisium madagascariensis DC	Apocynaceae				+	+
Triticum aestivum L.	Poaceae	Bread (processed wheat)			+	+
Vernonia amygdalina Delile	Asteraceae	Bitter leaf		Oriwo	+	+
Vigna unguiculata L.	Fabaceae	Cow pea		Ere	+	+
Vitellaria paradoxa Gaertn	Sapotaceae	Shea butter			+	+
Xylopia aethiopica (Dunal) A. Rich	Lauraceae			Unie	–	–
Zea mays L.	Poaceae	Maize	Corn flour	Okha	+	+
Zingiber officinale Roscoe	Zingibeaceae	Ginger			+	+

+ = Present; – = Absent

		Peri-urban market				Rural market			
Oba	Oregbene	Evbuotubu	Ugbogiobo	Ugbiohioko	Iguobazuwa	Ehor	Usen	Ekiadolor	Ugbogui
+	+	+	+	+	+	+	+	+	+
+	+	+	+	+	+	+	+	+	+
+	−	+	−	−	−	−	−	−	+
−	−	+	+	+	+	+	+	+	+
+	+	+	+	+	+	+	+	+	+
+	−	−	−	+	+	+	−	−	+
+	+	+	+	+	+	+	−	+	+
−	−	−	−	−	−	−	+	+	+
+	−	−	−	−	−	+	−	+	−
+	+	+	+	+	+	+	+	+	+
+	+	+	+	+	+	+	+	+	+
+	+	+	+	+	+	+	+	+	+
+	+	+	+	+	+	+	+	+	+
−	−	−	+	+	+	+	+	+	−
+	+	+	+	+	+	+	+	+	+
+	+	+	+	+	+	+	+	+	+

were more common in urban markets than in rural markets. On the other hand, *Tetrochidium didymostemon, Xylopia aethiopica, Anacardium occidentalis, Bombax buonopozense,* and *Dacryodes edulis* were found mostly in rural markets. However, this trend might not translate directly into plant diversity in urban and rural centers but their utilization patterns. In Benin City, peri-urban markets mostly act as a transition zone for rural and urban markets. Previous investigations of rural markets in Nigeria by Johnson and Johnson (1976) recorded 58 species of plants sold in Nigeria, Keratela and Hussain (1990) reported 21 species, Gill et al. (1993) recorded 93 plants, Idu et al. (2010) reported 103 and Osawaru and Odin (2012) reported 117. The difference in the number of plants recorded from the different study might be related to the season when the study was undertaken as well as the agricultural yield of the previous seasons and change in attitude, taste, where vendors source their plants, prevailing economic and environmental conditions. Mekasha and Tirfe (2019) highlighted that the marketing of agricultural produce requires planned production, grading of products, transportation to markets, distribution, pricing, and advertisement. Most of the plant species found in the markets are exotic. This supports the report of Muhanji et al. (2011) and Ogwu et al. (2016, 2017) which opined that the colonial era introduced and promoted the production and sale of plants exotic to Africa. Overall, the plants and plant products distribution across the market ranged from 58 to 83 in all the markets assessed (Fig. 1). There are more plants and plant products in urban markets compared to peri-urban and rural markets. This might be due to the higher food demands of the growing urban population, urbanization policies, greater economic power, and migration (Romanik 2008; Ogwu 2019). If the current urbanization trend is left unchecked, it might increase the vulnerability of African cities to climate change as well as challenges associated with food security. Another reason for the high distribution of plant species in urban markets might be the large size of these markets and the age-long attitude of rural dwellers to bring their farm produce to city centers for sale. The least diversity was in Usen Market while Uselu Market had the highest species composition. Overall, the urban markets had higher species composition. Next to the urban markets were peri-urban markets in plant and plant product abundance.

Plant products found in the different markets were assessed based on the level of processing that has been done to the plants. The classification of plant processing included first, second, and third-degree of plant and plant products (Fig. 2). The first degree of plant and plant products refers to the whole plant or plant part, while the second degree of plant and plant products and third degree of plant and plant products are processed and reprocessed plant or plant parts, respectively. It was also observed that most plant products were only processed once before been presented for sale in the markets (Fig. 2).

Plants and plant products in the local markets can be grouped into cereals (e.g., maize, rice, guinea corn, millet etc.), legumes (e.g., beans, groundnut, soybean etc.), stem tubers (e.g., yam and Irish potato), root tubers (e.g., cassava, carrot and sweet potato), fruits (e.g., pawpaw, orange, pineapple, mango, banana, pear, etc.), vegetables (e.g., waterleaf, bitter leaf, *Amaranthus* sp., *Celosia* sp., pumpkin leaf etc.), nuts (e.g., coconut), oil (e.g., palm fruit), spices (e.g., pepper, onion, ginger, garlic etc.).

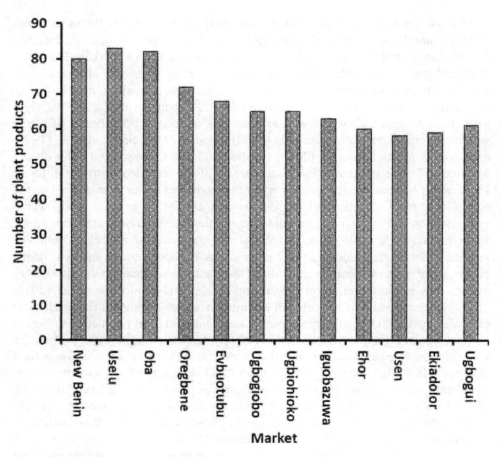

Fig. 1 Number of plants and plant products in local markets in Benin City and environs

Fig. 2 Category of plants
based on forms in which they
are available in the market

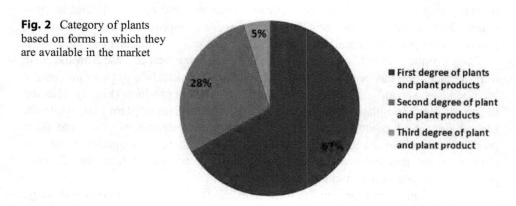

The dominance of first degree processed plant products could be attributed to the
near lack of government support for the agriculture sector of Nigeria. It also suggests
Nigerians might prefer plant products that have undergone little to no processing. It

was observed that the availability of plants and their products rely on seasonal variations. The plants and plant products were seen in different forms in the markets depending on the season. Markets are made up 90–99% of plants and plant products. Some of these products may have been to the 3rd degree of processing, for example, Rain boot, etc. However, not all part of the plant may be essential. These forms are fruits, leaves, rhizomes, bulbs, corms, stem tuber, root tuber and also in processed forms such as fufu, gari, and cassava flour from cassava; oil, broom, and basket from oil palm; tomato paste from tomato, etc. These different forms are the status in which these plants and plant products are best sold and preserved. This is in line with the report of Idu et al. (2005).

The utilization pattern of the plants and plant products is presented in Table 3. Major categories included cereals, legumes, roots and tubers, fruits, fats and oils, sugar crops, fiber crops, spice and condiments, beverages, medicinal, and others. It was observed that 86% of the plants and plant products are used for foods, 10% are used for medicinal purpose while 3% is used for other purposes. However, the highest percentage was noticed in fruits, which are 35%.

The habit of plants found in local markets in Benin City and environs range from grasses, herbs, shrubs, and trees to vines. Overall, the composition was 39%, 20%, 17%, 15%, and 9% for trees, shrubs, grasses, herbs, and vines, respectively (Fig. 3). This is an indication that for plants and plant products gotten from tree crops, buyers, and vendors will have to wait for months or years before parts that are available to be harvested for sale or consumption.

The plant species recorded in the study are mainly used as a source of food, cash or medicine. Others supply diversity, essential nutrients, vitamins, or minerals in diets that would otherwise consist primarily of carbohydrates (Johns 2004; Johns and Sthapit 2004). Our investigation revealed that most of the plants and plant products are mainly for foods while a few are for medicinal and other uses. Often, they reflect cultural values and as a pool of health and nutritional information for the public and health practitioners (Johns and Eyzaguirre 2002). The availability of diverse plant and plant products in the market relies on local agriculture system, seasonal variations, local knowledge and practices, as well as plant germplasm. Produce are mainly sourced locally from home gardens and distant farms. Therefore, the vulnerability of agriculture systems in Benin City to climate change might not be affecting the production of the plants reported in this chapter in the short term. However, the interplay of diverse external factors and climate change is likely to affect the availability of plant and plant products currently found in open markets in the long term. It is recommended that the roles of women, soil health, and plant diversity be assessed in order to formulate a policy to mitigate and adapt to climate change impact on agriculture and food security in Benin City and other parts of Nigeria and Africa.

Moreover, considering that Muhanji et al. (2011) reported that there might be 45,000 plant species in Africa, the amount of plant and plant products recorded in open markets in Benin City only represent a small portion of that diversity. About 86% of the plants and plant products found in the markets in Benin City and environs are used for foods, 11% are for medicinal purpose, while 3% are used for others

Table 3 Utilization pattern of plants and plant products in local markets within Benin City and environs

Botanical name	Common name	Cereal	Legumes	Root and tubers	Vegetables	Fruits and nuts	Fats and oil	Sugar crop	Fiber crop	Condiments and spices	Beverage and stimulant	Corm and rhizome	Medicinal	Other
Abehmoshus esculentus	Okra				•	•								
Adensonia digitata	Baobab					•								
Allium cepa	Onion										•			
Allium sativum	Garlic										•		•	
Amaranthus caudatus	Spinach				•									
Anacardium occidentalis	Cashew					•								
Ananas comosus	Pineapple					•								
Annona mutricata	Sour sop					•								
Arachis hypogeaa	Groundnut		•											
Azadirachta indica	Dogoyaro												•	
Bombax buonopozense						•								
Brassica oleracea	Cabbage				•									
Calotropis procera										•				•
Capsicum annum	Pepper					•								
Capsicum frutescens	Pepper					•								
Carica papaya	Pawpaw					•								
Celosia argentea	Celosia				•									
Citrus aurantifolia	Lime					•							•	
Citrus limon	Lemon					•							•	
Citrus sinensis	Orange					•								

(continued)

Table 3 (continued)

Botanical name	Common name	Cereal	Legumes	Root and tubers	Vegetables	Fruits and nuts	Fats and oil	Sugar crop	Fiber crop	Condiments and spices	Beverage and stimulant	Corm and rhizome	Medicinal	Other
Cochorus olitoriu	Jute				•				•					
Cocos nucifera	Coconut					•								
Cocus nucifera	Vegetable oil						•							
Cola acuminate	Kolanut					•								
Cola nitida	Kolanut					•								
Colocasia esculenta	Cocoyam											•		
Crescentia cujele	Calabash													•
Cucumeropsis mannii	Melon		•											
Cucumis sativus	Cucumber					•								
Cucurbita pepo						•								
Cucurma longa										•				
Cymbopogon citrates	Lemmon grass												•	
Cyperus esculentus	Tiger nut											•	•	
Dacryodes edulis	African pear					•								
Daucus carota	Carrot			•	•									
Dennettia tripetala	Pepper fruit					•								
Dialium guineense						•								
Dioscorea alata	Water yam			•										
Dioscorea rotundata	Yam			•										

Scientific name	Common name
Elaeis guineensis	Palm oil
Garcina cola	Bitter cola
Glycine max	Soy bean
Gnetum africanum	
Gossypium hirsutum	Cotton
Hibiscus cannabinus	
Hibiscus sabdarifa	Roselle
Ipomea batata	Sweet potato
Irvingia gabonensis	Bush mango
Lycopersicum esculentum	Tomato
Mangifera indica	Mango
Manihot esculatua	Cassava
Murraya koenigii	Curry
Musa paradisiaca	Banana
Musa sapientum	Plantain
Myristica fragrans	Nutmeg
Ocimum gratissimum	Scent leaf
Oryza sativa	Rice
Parkia clappertorniana	Locust bean
Pennisitum glaucum	Millet

(continued)

Table 3 (continued)

Botanical name	Common name	Cereal	Legumes	Root and tubers	Vegetables	Fruits and nuts	Fats and oil	Sugar crop	Fiber crop	Condiments and spices	Beverage and stimulant	Corm and rhizome	Medicinal	Other
Pentaclethra macrophylla	African oil bean		•											
Persea americana	Avocado Pear					•								
Phaseolus vulgaris	Beans		•											
Phoenix dactylifera	Date palm					•								
Piper guineense	African pepper					•								
Psidium guajava	Guava					•								
Rauwolfia vomitoria	Rauwolfia												•	
Ricinus communis	Castor oil		•											
Saccharum officinarum	Sugar cane							•						
Sesamum orientale						•								
Solanum melogena	Garden-egg					•								
Solanum tuberosum	Irish potato			•										
Sorghum bicolour	Guinea corn	•												
Spondias mombin	Hug phem					•								
Talfairia occidentalis	Pumpkin				•								•	
Talinum triangulare	Water leaf				•									

Tamarindus indica L.						•			•
Tetrochidium didymostemon					•				•
Thaumatococcus danielli			•						
Theobroma cacao — Cocoa				•			•		
Thymus vulgaris — Thyme				•		•		•	
Treculia africana					•			•	
Trilepisium madagascariensis						•			
Triticum aestivum — Wheat	•								
Vernonia amygdalina — Bitter leaf						•			
Vigna unguiculata — Cow pea		•							
Vitellaria paradoxa — Shea butter						•	•		
Xylopia aethiopica							•		
Zea mays — Maize		•							
Zingiber officinale — Ginger			•					•	•

Fig. 3 The habit of plants and plant products in local markets within Benin City and environs

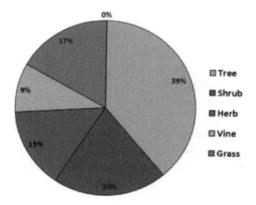

purpose. Of the 86% used for food purpose, fruits and vegetable had the highest percentage of usage, which are 35% and 9% for fruits and vegetables, respectively. This is an indication that the markets in Benin City and environs have rich and diverse pool of fruits and vegetables, which are of great nutritional value. This is in line with the study of Odhav et al. (2007) and Ogwu et al. (2016) wherein they pointed out that indigenous vegetables and fruits represent inexpensive but high quality nutrition sources for the poor segment of the population. Since many indigenous food plants grow wild, they are accessible, they can be collected freely and are thus available to everyone, including the poor (Kabuye et al. 1999). Fruits and vegetables are of great nutritional value. They are important sources of vitamins and minerals that are essential for human health and well-being. Their consumption ensures the intake of various essential vitamins and mineral elements thus avoiding the problem of malnutrition (Yamaguchi 1983). There is a wide variety of indigenous vegetables and fruits found in Africa, which are chief sources of nutrients, vitamins, antioxidants, minerals, and proteins (Odhav et al. 2007; Ogwu et al. 2016; Ogwu 2020). Some of the indigenous vegetables and fruits are mainly used for medicinal purposes (Eifediyi et al. 2008).

Names and naming are important determinant factors in local society and contributing to promoting sustainable plant utilization and conservation (Penny 2001; Ogwu and Osawaru 2014; Ogwu et al. 2014). Plants are more easily recognized by their local names in every part of the world. These local names play a vital role in ethnobotanical study of a specific tribe or region (Singh 2008). The local names of crop plants, especially in Bini language, are reported among tribes surveyed. Documentation of local names is highly valued by Rogers (1963), Rogers and Applan (1973), Allem (2000), Bressan et al. (2005), Sawadogo et al. (2005), and Osawaru and Dania-Ogbe (2010). Local names are used to promote and trade plant and plant products in all the local markets assessed with little to no reference to their scientific nomenclature. Although local names are not directly recommended for scientific discussions because they lack uniformity and consistency (Singh 2008), yet they may certainly be considered as a useful tool for obtaining useful information on plants. Local names provide means of reference by local people in a particular area. Also, in some cases, the plants are well known with their local name than the common names. This is the case of "ebolebo" and "dogoyaro" for Indian almond and neem, respectively.

In conclusion, plants provide valuable functions as foods, raw materials, socio-economic development, as well as sustainable environmental development and indicator of climate change impacts. They have been used as a means of livelihood sustenance and preservation of indigenous knowledge through their utilization pattern for centuries. This chapter established that local markets are a data bank for economic plant species. The diversity of these plants and plant products in the various local market assessed suggest that despite ongoing climate change, some level of plant production and plant-human interaction is ongoing. This chapter also revealed that urban markets have a higher number of plants and plant products compared to peri-urban and rural markets. This is a reflection that lesser populated markets are less diverse in terms of plants and plant products in the market. Additionally, majority of the plants and plant products are utilized as food with fruits being the predominant part that is used as food. Overall, the vendors/sellers are predominantly women and their interaction with plants in the market makes them an important group in the fight against climate change, food insecurity, and biodiversity crisis.

References

Agea GJ, Kimondo JM, Okia CK, Abohassan RAA, Obua J, Hall J, Teklehaimanot Z (2011) Contribution of wild and semi wild food plants to overall household diet in Bunyoro Kitara Kingdom. Ugandan Agric J 6(4):134–144

Allem AC (2000) Ethnobotanical testimony on the ancestors of cassava (*Manihot esculenta* Crantz. subsp. *esculenta*). Plant Genet Resour Newsl 123:19–22

Atedhor GO, Odjugo PAO, Uriri AE (2011) Changing rainfall and anthropogenic-induced flooding: impacts and adaptation strategies in Benin City, Nigeria. J Geogr Reg Plan 4(1):42–52

Bressan EA, Veasey EA, Peroni N, Felipim AP, Pacheo dos Santos KM (2005) Collecting yam (*Dioscorea* spp.) and sweet potato (*Ipomea batatas*) germplasm in traditional agriculture small-holding in the Vale do Riberira, Sao Paulo, Brazil. Plant Genet Resour Newsl 144:8–13

Burkill RM (1985) The useful plants of tropical West Africa, vol 1. Royal Botanic Gardens, Kew. 635p

Bye RA, Linares E (1983) The role of plants found in the Mexican markets and their importance in ethnobotanical studies. J Ethnobiol 3:1–13

Cunningham AB (2001) Applied ethnobotany: people, wild plant use and conservation. People and plants conservation manual. Earthscan, London. 300p

De Caluwe E (2011) Market chain analysis of baobab (*Adansonia digitata* L.) and tamarind (*Tamarindus indica* L.) products in Mali and Benin. PhD thesis, Faculty of Bioscience Engineering, Ghent University, Ghent

Efe SI, Eyefia OA (2014) Urban warming in Benin City, Nigeria. Atmos Clim Sci 4:241–252

Eifediyi K, Mensah JK, Ohaju-Obodo JO, Okoli RI (2008) Phytochemical, nutritional and medic-inal properties of some leafy vegetables consumed by Edo people of Nigeria. Afr J Biotechnol 7:2304–2309

Floyd AC, Oikpor R, Ekene B (2016) An assessment of climate change in Benin City, Edo State, Nigeria. FUTY J Environ 10(1):87–94

Gill LS, Nyawuame HGK, Omoigui JD (1993) A survey of economic plants sold in the market of Benin City (Edo State) Nigeria. Niger Technol Educ 10:69–77

Idu M, Osawaru ME, Orhue E (2005) Ethno-medicinal plant products in some local markets on Benin City, Nigeria. Ethnobotany 17:118–122

Idu M, Erhabor JO, Efijuemue HM (2010) Documentation on medicinal plants sold in markets in Abeokuta, Nigeria. Trop J Pharm Res 9(2):110–118

Johns T (2004) Underutilized species and new challenges in global health. LEIZ Mag 20(1):5–6

Johns T, Eyzaguirre PB (2002) Nutrition and the environment. In: Nutrition: a foundation for development, vol 20. ACC/SCN, Geneva, pp 269–285

Johns T, Sthapit BR (2004) Biocultural diversity in the sustainability of developing country food systems. Food Nutr Bull 25(2):143–155

Johnson EJ, Johnson TJ (1976) Economic plants in a rural Nigerian market. Econ Bot 30:375–381

Kabuye CHS, Maundu PM, Ngugi W (1999) Traditional food plants of Kenya. Kenya Resource Centre for Indigenous Knowledge, Nairobi

Keratela YY, Hussain HSH (1990) Indigenous fruits sold in markets of Ishan area of Bendel State. The Nigerian Field 55:12–18

Martin G (1992) Searching for plants in peasant marketplaces. In: Plotkin MJ, Famolare L (eds) Sustainable harvest and marketing of rainforest products. Island Press, Washington, DC, pp 212–223

Mekasha J, Tirfe GG (2019) Assessing factors affecting marketing of vegetable products: the case of Qewet woreda, Ethiopia. IOSR J Bus Manag 21(4):82–93

Muhanji G, Roothaert R, Webo C, Stanley M (2011) African indigenous vegetable enterprise and market access for small-scale farmers in East Africa. Int J Agric Sustain 9(1):194–202

Nguyen MT (2005) Cultivated plant collections from marketplaces. Ethnobot Res Appl 3:5–15

Obiri BD, Addai A (2007) People and plants: a survey of economic botanicals on the Kumasi central market. Ghana J For 21 and 22:50–71

Odhav B, Beekrum S, Akula U, Baijnath H (2007) Preliminary assessment of nutritional value of traditional leafy vegetables in KwaZulu-Natal, South Africa. J Food Compos Anal 20:430–435

Ogwu MC (2019) Towards sustainable development in Africa: the challenge of urbanization and climate change adaptation. In: Cobbinah PB, Addaney M (eds) The geography of climate change adaptation in urban Africa. Springer Nature, Cham. 29–55pp. https://doi.org/10.1007/978-3-030-04873-0_2

Ogwu MC (2020) Value of *Amaranthus* [L.] species in Nigeria. In: Waisundara V (ed) Nutritional Value of Amaranth. IntechOpen, London. 1–21pp. https://doi.org/10.5772/intechopen.86990

Ogwu MC, Osawaru, ME (2014) Comparative Study of Microflora Population on the Phylloplane of Common Okra [Abelmoschus esculentus L. (Moench.)]. Nigerian Journal of Biotechnology 28:17–25

Ogwu MC, Osawaru ME, Ahana CM (2014) Challenges in conserving and utilizing plant genetic resources (PGR). Int J Genet Mol Biol 6(2):16–22. https://doi.org/10.5897/IJGMB2013.0083

Ogwu MC, Osawaru ME, Aiwansoba RO, Iroh RN (2016) Status and prospects of vegetables in Africa. In: Borokini IT, Babalola FD (eds) Conference proceedings of the joint biodiversity conservation conference of Nigeria Tropical Biology Association and Nigeria Chapter of Society for Conservation Biology on MDGs to SDGs: toward sustainable biodiversity conservation in Nigeria. University of Ilorin, Ilorin. 47–57pp

Ogwu MC, Osawaru ME, Obahiagbon GE (2017) Ethnobotanical survey of medicinal plants used for traditional reproductive care by Usen people of Edo State, Nigeria. Malaya J Biosci 4(1):17–29

Osawaru ME, Dania-Ogbe FM (2010) Enthnobotanical revelations and traditional uses of West African Okra [*Abelmoschus caillei* (A. Chev.) Stevels] among tribes in South-Western Nigeria. Plant Arch 10:211–217

Osawaru ME, Odin EI (2012) An inventory of plants and plant products in rural and urban markets in some localities in Southern Edo State, Nigeria. Univ Benin J Sci Technol 1:1–13

Osawaru ME, Ogwu MC (2014a) Conservation and utilization of plant genetic resources. In: Omokhafe K, Odewale J (eds) Proceedings of 38th annual conference of The Genetics Society of Nigeria. Empress Prints Nigeria Limited, Benin City, Nigeria, pp 105–119

Osawaru ME, Ogwu MC (2014b) Ethnobotany and germplasm collection of two genera of cocoyam (*Colocasia* [Schott] and *Xanthosoma* [Schott], Araceae) in Edo State Nigeria. Sci Technol Arts Res J 3(3):23–28. https://doi.org/10.4314/star.v3i3.4

Penny RA (2001) Gender and Indigenous knowledge experiences in Nigeria and the USA. Indig Knowl Dev Monit 9(1):16–17

Rogers DJ (1963) Studies of *Manihotesculenta* Cranz and related species. Bull Torrey Bot Club 90:43–54

Rogers DJ, Applan SG (1973) *Manihot and Manihotoides* (Euphorbiaceae), a computer assisted study, Flora Neotropica, monograph no 13. Hofner Press, New York

Romanik C (2008) An urban-rural focus on food markets in Africa. Research report. The Urban Institute. 45p. https://www.urban.org/sites/default/files/publication/31436/411604-An-Urban-Rural-Focus-on-Food-Markets-in-Africa.PDF

Sawadogo M, Duedraogo J, Belem Balma M, Dossou B, Jarvis D (2005) Influence of ecosystem component on cultural practices affecting the in situ conservation of agricultural biodiversity. Plant Genet Resour Newsl 141:19–25

Singh H (2008) Importance of local names of some useful plants in ethnobotanical study. Indian J Tradit Knowl 7(2):365–370

Whitaker TW, Cutler HC (1966) Food plants in a Mexican market. Econ Bot 20:6–16

Yamaguchi M (1983) World vegetables, principle, production and nutritive values. Ellishorwood Limited Publishers, Chichester

Yaradua SS, El-Ghani MA (2015) Ethnobotanical survey of edible plants sold in Katsina metropolis markets. Int J Sci Res 4:884–889

Permissions

All chapters in this book were first published by Springer; hereby published with permission under the Creative Commons Attribution License or equivalent. Every chapter published in this book has been scrutinized by our experts. Their significance has been extensively debated. The topics covered herein carry significant findings which will fuel the growth of the discipline. They may even be implemented as practical applications or may be referred to as a beginning point for another development.

The contributors of this book come from diverse backgrounds, making this book a truly international effort. This book will bring forth new frontiers with its revolutionizing research information and detailed analysis of the nascent developments around the world.

We would like to thank all the contributing authors for lending their expertise to make the book truly unique. They have played a crucial role in the development of this book. Without their invaluable contributions this book wouldn't have been possible. They have made vital efforts to compile up to date information on the varied aspects of this subject to make this book a valuable addition to the collection of many professionals and students.

This book was conceptualized with the vision of imparting up-to-date information and advanced data in this field. To ensure the same, a matchless editorial board was set up. Every individual on the board went through rigorous rounds of assessment to prove their worth. After which they invested a large part of their time researching and compiling the most relevant data for our readers.

The editorial board has been involved in producing this book since its inception. They have spent rigorous hours researching and exploring the diverse topics which have resulted in the successful publishing of this book. They have passed on their knowledge of decades through this book. To expedite this challenging task, the publisher supported the team at every step. A small team of assistant editors was also appointed to further simplify the editing procedure and attain best results for the readers.

Apart from the editorial board, the designing team has also invested a significant amount of their time in understanding the subject and creating the most relevant covers. They scrutinized every image to scout for the most suitable representation of the subject and create an appropriate cover for the book.

The publishing team has been an ardent support to the editorial, designing and production team. Their endless efforts to recruit the best for this project, has resulted in the accomplishment of this book. They are a veteran in the field of academics and their pool of knowledge is as vast as their experience in printing. Their expertise and guidance has proved useful at every step. Their uncompromising quality standards have made this book an exceptional effort. Their encouragement from time to time has been an inspiration for everyone.

The publisher and the editorial board hope that this book will prove to be a valuable piece of knowledge for researchers, students, practitioners and scholars across the globe.

List of Contributors

Godfrey Odongtoo
Department of Computer Engineering, Busitema University, Tororo, Uganda
Department of Information Technology, Makerere University, Kampala, Uganda

Denis Ssebuggwawo
Department of Computer Science, Kyambogo University, Kampala, Uganda

Peter Okidi Lating
Department of Electrical and Computer Engineering, Makerere University, Kampala, Uganda

Salé Abou and Madi Ali
National Advanced School of Engineering of Maroua (ENSPM), The University of Maroua, Maroua, Cameroon

Anselme Wakponou
Faculty of Arts, Letters and Human Sciences (FALSH), The University of Ngaoundéré, Ngaoundéré, Cameroon

Armel Sambo
Faculty of Arts, Letters and Human Sciences (FALSH), The University of Maroua, Maroua, Cameroon

Abubakar Hamid Danlami
Department of Economics, Faculty of Social Sciences, Bayero University Kano, Kano, Nigeria

Shri Dewi Applanaidu
Department of Economics and Agribusiness, School of Economics, Finance and Banking, College of Business, Universiti Utara Malaysia, Sintok, Malaysia

Argaw Tesfaye
Department of Geography and Environmental Studies, Mekdela Amba University, Mekane Selam, Ethiopia

Arragaw Alemayehu
Department of Geography and Environmental Studies, Debre Berhan University, Debre Berhan, Ethiopia

David O. Chiawo
Strathmore University, Nairobi, Kenya

Verrah A. Otiende
Pan African University Institute for Basic Sciences Technology and Innovation, Nairobi, Kenya

Mike Muller
Wits School of Governance, University of the Witwatersrand, Johannesburg, South Africa

B. E. Fawole
Department of Agricultural Extension and Rural Development, Federal University, Dutsinma, Nigeria

S. A. Aderinoye-Abdulwahab
Department of Agricultural Extension and Rural Development, Faculty of Agriculture, University of Ilorin, Ilorin, Nigeria

Chukwudi Nwaogu
Department of Forest Protection and Entomology, Faculty of Forestry and Wood Sciences, Czech University of Life Sciences, Prague 6-Suchdol, Czech Republic
Department of Environmental Management, Federal University of Technology, Owerri, Nigeria
Department of Ecology, Faculty of Environmental Sciences, Czech University of Life Sciences, Prague 6-Suchdol, Czech Republic

Edward M. Mungai and S. Wagura Ndiritu
Strathmore University Business School, Nairobi, Kenya

Izael da Silva
Strathmore University, Nairobi, Kenya

M. R. Motsholapheko
Water Resources Management Program, Okavango Research Institute, University of Botswana, Maun, Botswana

B. N. Ngwenya
Ecosystems Services Program, Okavango Research Institute, University of Botswana, Maun, Botswana

Ugbah Paul Akeh and Olumide A. Olaniyan
National Weather Forecasting and Climate Research Centre, Nigerian Meteorological Agency, Abuja, Nigeria

Steve Woolnough
Department of Meteorology, University of Reading, Reading, UK

Moses Edwin Osawaru
Department of Plant Biology and Biotechnology, Faculty of Life Sciences, University of Benin, Benin City, Edo State, Nigeria

Matthew Chidozie Ogwu
Department of Plant Biology and Biotechnology, Faculty of Life Sciences, University of Benin, Benin City, Edo State, Nigeria
Scuola di Bioscienze e Medicina Veterinaria, Università di Camerino – Centro Ricerche Floristiche dell'Appennino, Parco Nazionale del Gran Sasso e Monti della Laga, Barisciano (L'Aquila), Italy

Index

9 781641 167352